Changing Works

DOUGLAS HARPER

Changing Works

VISIONS OF A LOST AGRICULTURE

The University of Chicago Press — Chicago and London

DOUGLAS HARPER is professor of sociology at Duquesne University and the author of *Good Company* and *Working Knowledge: Skill and Community in a Small Shop.*

The University of Chicago Press, Chicago 60637
The University of Chicago Press, Ltd., London

Printed in Hong Kong

10 09 08 07 06 05 04 03 02 01 1 2 3 4 5
ISBN: 0-226-31722-6 (cloth)

Library of Congress Cataloging-in-Publication Data
Harper, Douglas A.
Changing works : visions of a lost agriculture / Douglas Harper.
p. cm.
Includes bibliographical references and index.
ISBN 0-226-31722-6 (cloth : acid-free paper)
1. Agriculture–United States. 2. Agricultural innovations—United States. 3. Dairy farming–New York (State) 4. Agriculture–United States—Pictorial works. 5. Agricultural innovations–United States—Pictorial works. 6. Dairy farming—New York (State)—Pictorial works. I. Title.
S441 .H2943 2001
630'.973'022–dc21 00-011318

♾ The paper used in this publication meets the minimum requirements of the American National Standard for Information Sciences—Permanence of Paper for Printed Library Materials, ANSI Z39.48-1992.

To the memory of my father, and to Colter and Molly, who have North Country roots. And to farmers who know animals and land and know how to make a system that works for them and for us as well.

Contents

Acknowledgments

A book this long in the making accumulates many debts. Forty-eight farmers invited me to their farms for interviews that lasted either several hours or several days. My thanks to them!

Special thanks to Art Henry, Willie Louie, Mason O'Brien, Rob Evans, Kevin, Marian, and the rest of the Acres family, Jim and Emily Fisher, Tom and Marie Lee, and Jim Carr for interviews extending over hours or several days. Karen Kline, research assistant and friend, helped with interviews, analysis, and years of encouragement. To these North Country friends, my thanks. Several of these people are no longer with us, and I hope, with this book, to honor their memory.

At Cornell's Department of Rural Sociology, Tom Lyson and Gil Gillespie set the stage for this project in their own research, and I hope they see this as a continuation of their good work. Their help has been critical for more than a decade, and as a result I've become a little more of a quantoid. Not terribly so, but the lessons haven't hurt.

This project extended over my tenure at three universities. At SUNY Potsdam, a faculty development grant paid for the aerial photography, and a yearlong sabbatical in 1989–90 offered by President Humphrey Tonkin supported the initial data gathering. My Potsdam colleagues Harry Kienzle, Walter Weitzmann, and Bill Mathews helped me maintain the idea that the project was a worthwhile endeavor. At Potsdam there was a strong interest in local research and many small ways of helping it happen, despite the crushing teaching loads of that system that so stifle research. I was able to keep the project on life support (but little more) during four years swimming with the gators as chair of sociology at the University of South

Florida. In my current position at Duquesne University, where I am also chair but sans gators, I have been able to do the lion's share of the writing. This is, in large part, due to a congenial department and support from my dean, Constance Ramirez, and Provost Michael Weber. Both Doctors Ramirez and Weber have offered material and psychological support, and my gratitude to them is extreme.

In my current department, Chuck Hanna read two drafts of the manuscript and offered detailed and useful suggestions. My office staff, headed by Cheri Cunningham, organized the last stages of the manuscript preparation with the greatest of skill and good will. My colleagues in the Center for Interpretive and Qualitative Research at Duquesne have also guided final revisions. Thanks to all!

I've offered several papers from this research to the visual sociology community. Colleagues who listened, looked, and commented include Rob Boonzajer Flaes and Ton Guiking of the University of Amsterdam, Patrizia Faccioli and Pino Losacco of the University of Bologna, Ricabeth Steiger of the Swiss National Museum of Photography, Jon Prosser of Leeds University, Elizabeth Edwards of the Open University, and Chuck Suchar of DePaul University. Readers who made critical recommendations were Kathryn Dudley and Howard S. Becker. Kathryn Dudley helped me more fully develop issues pertaining to domestic and productive technology, and Howie Becker continues to be a mentor who knows perfectly the balance between just criticism and hopeful encouragement. Always ironic; never sarcastic. To both I owe a great debt.

At the University of Louisville my work in the SONJ archive was ably assisted by James Anderson and Bill Carner. During my Louisville research I was hosted by my friend and colleague Jon Rieger, who, in the meantime, taught me a lot of rural sociology and photography.

Two old friends who grew up on farms, Rob Hawkinson and Ed Lafontaine, helped me understand how the issues described in this book influenced the lives of their parents and neighbors. Heidi Larson, who is not a farmer, reminded me continually about the moral and ethical aspects of our relationships with cows.

For the third time at the University of Chicago Press I've been able to work with Doug Mitchell, an editor who takes risks and treats authors well. His insight is as sharp as a razor and, at times, as intimidating. Robert Devens assisted mightily. Copy editor Erik Carlson went beyond copy editing to clarification. The Press welcomed my involvement in the design of the book, and in this case the design is the dancer and the book is the dance. Their accommodation of my desires has been extraordinary. I acknowledge in particular the artful work of designer Michael Brehm.

I'd like to thank some "unmet mentors." For this project, they include Sally McMurry, Craig Canine, Verlyn Klinkenborg, Katherine Jellison, and Steven Plattner. Many other writers and researchers were, of course, indispensable, but these deserve special mention. It is rather nice to be able to thank folks I've never met, so thanks for help you didn't know you provided!

Finally, I thank my family. My immediate family—Suzan, Molly, and Colter—lived through the early stages of this research. They made me hang my barn clothes on the front porch even when it was twenty below zero but otherwise were great supporters. When the project started we were North Country people; now we are Pittsburghers, and Molly has gone west. Maybe that is the good part of a long project: you see your family's growth in the memories of all the work. Suzan provided a substantial bedrock of support throughout.

My mother and father, Herb and Norma Harper, read several versions of the book. As in the case of everything I have written, they were sharp, unsparing, and enormously supportive editors.

My father approached the end of his life as the book neared completion, and in the last days his handwriting on the pages lost its elegance. One of his master's theses was on soil erosion and farming techniques in the 1950s, so this project is an homage to a man I was lucky to be the son of. The raging loneliness brought by his departure is to some extent mitigated by the sense of carrying forward one of his life themes.

Changing Works

Introduction

Fig. 1 *(previous pages)* Spreading manure in a field on the Deck farm on U.S. 22, west of Allentown, Pennsylvania, May 1945. Photograph by Sol Libsohn. *Note:* The dates assigned to photographs in the Standard Oil of New Jersey archive may represent the dates on which they were cataloged, which may be one or two months later than the dates on which the photographs were taken.

I moved to a farmhouse in New York's North Country in the mid-1970s.[1] It was several miles and several cultures from a college in a neighboring town, where I taught. I would hear the house creak and groan in the autumn winds, and I sensed farmers' spirits, heading in after the last corn was harvested. Several times, alone in the house, I opened the back door to the woodroom, startled by an unusual sound, sure someone was there. It was no different in the barns. Dusty cobwebs covered the four horse stalls, which had held animals that had powered farming for a hundred years. Leather harnesses hung against the wall, black with horse sweat and hardened from years of neglect. The floor of the barn was worn in gentle paths, and, as I stood quietly in the empty barn, I imagined the snorts of horses, the sounds of hooves scraping on the floor and the "Whoa there!" of farmers readying them for work. Who were those people? What had been their work? How were the farmers I watched crossing the fields atop their hundred-horsepower tractors connected to the traditions of the empty barn?

I was a stranger to my new neighbors, and the walls between us broke down slowly. I shared the farmhouse with my colleague and friend Ed, and on one of the first weekends in our first fall we approached our neighbor, Carmen Acres, to borrow equipment to cut and haul wood. We had installed a woodstove in the farmhouse, and Carmen, who was keeping heifers in our barn, loaned us an old tractor and manure spreader to use as a trailer. Carmen told us to help ourselves to the dead elm trees on the field edges, alongside fences made from the stones cleared a hundred years before. Carmen seemed bemused that we were going to heat the big farmhouse with wood, but he was pleased to have dead fence-row elm cut from his field's edges.

Carmen's generosity seemed natural in the setting of the neighborhood. Farmers had worked with their neighbors, in one way or another, since the beginning of the neighborhood. Ed, who had grown up on a farm in Ohio, accepted this naturally. I had lived in cities for the previous ten years, and I was caught off guard by the casual generosity of my new neighbor.

I lived in this neighborhood for sixteen years. I began my family in the farmhouse, and my children spent their early years immersed in the surrounding rural culture. My wife, Suzan, became a farmer in her own right when she established a greenhouse business in solar buildings we built from lumber taken from Carmen's trees.

Much seemed right in this world. The patchwork of fields tucked into woods and swamps was farmed in a way suited to the environment. Though the ground was rocky and the weather cold, the land was fertile. Cow manure was a nutrient rather than a nuisance or even a hazardous waste, as it was becoming on the factory farms of the south.

Families had farmed the neighboring land for generations, adapting to evolving agricultural technologies. The farms were small enough that a family, with a little help, could usually manage on its own (which is not to say that they all managed well, or that family life on all farms was smooth or easy). The local economy was interconnected with the dairy industry, and it provided jobs and reasons for living in a northern rural neighborhood. People knew each other, and they had depended on each other for generations. Their mistakes as well as successes were on the table, and maybe people had a more realistic view of their potential—both good and bad—because of it. Mostly I valued the dairy community because it was based on work which required patience and imagination. The dairy farmers I came to know had the self-confidence carried by people who face physical environments successfully.

But much was also not so right in the neighborhood. The number of working farmers declined dramatically even during the years we lived there. I was myself occupying a farmhouse which had been productive for over a hundred years. The rural society was becoming fragmented and diluted—filled up with strangers seeking a pastoral respite or a cheap place to live. The farmers, ever fewer in number, farmed in increasing isolation from each other. As I watched them from the small country roads, they seemed like lonely masters of ever larger and more expensive machines.

My experiences in a changing neighborhood inspired this study. I came to see the issues as related to a simple theme: that is, the question of how people organize themselves socially and technically to make milk. The milk under examination, of course, comes from a cow, which runs machine-like, on a daily schedule. The milk flows because she has given birth, and it dries up when the natural cycle indicates that her calf would be self-sufficient. In the meantime the farmer has bred her again, making a milk machine for another year. Her calf is either raised to become a milk machine like her mother, or butchered for veal if unfortunate enough to be born a bull. At the basis of the flow of milk is an animal which operates as it has since its ancestors were first domesticated. What has changed is the way work is organized around the cow, and the machinery which has been developed to make the process more like mass production and less like a craft.

In the background is the realization that we apply expertise to make work easier or more profitable. The technologies that result may eliminate work and increase production, but they cost money, making larger farms necessary. The technology takes the farmer further away from the land, the animals, and other farmers with whom he or she had neighbored and worked. Technology introduces a rational calculus to farming, and traditional ways of thinking and doing give way to scientific methods and modern economic considerations. The damage done to communities, the land, and animals comes to be seen as an inevitable side effect of what comes to be viewed, strangely, as progress.

This book examines the transformation of farming during and just after World War II as a context for understanding modern agriculture. The World War II era brought rapid mechanization and the end of animal power and cooperative farming, which farmers called "changing works." As we live through the further developments of these revolutions, I believe we should question their implications for food production, rural communities, and the environment.

The Project: An Overview

Two springs after moving to the North Country, I worked for Carmen for six weeks, helping with the spring planting. I meant this as a combination of agricultural work and sociological observation, which the family seemed to accept (I was, after all, free labor). I drove tractors on small jobs, built a chimney, accompanied Carmen on trips to neighboring farms as he sold and delivered agricultural equipment, unloaded several tons of fertilizer and seed, ate with the family, overheard day-to-day conversations, and initiated my own. The immersion was nearly total, and I achieved it by simply driving around the block and taking whatever job I was handed.

I also completed several in-depth interviews with neighboring farmers during this period. These were histories of individual farms, focusing on the impact of evolving technology on the social organization of farming.

In the ensuing years my dairy farm project restarted and went on hold several times, as other projects nudged their way to the front of the queue. I devoted a sabbatical year (1989–90) to the study, interviewing nearly fifty farmers in the immediate neighborhood.[2] It was during that time that I began to work with Cornell University's Department of Rural Sociology, and I began to see my community in the context of changes occurring on a national and even international level.

In midstudy, my family and I moved from the North Country to a university several climate zones away. In the meantime we have moved again. Over these past nine years I have returned several times to the North Country to do more interviewing and data collection, the last time in 1999. I have been pained at the separation from our good life in the North Country but, perhaps, through my separation, more able to see the North Country farmer in the context of history, national policy, and the rural cultures in many parts of the world.

The Standard Oil of New Jersey Archive

One of the ways I studied how farming changed during the lifetime of my neighbors was through their interpretations of photographs made during the Standard Oil of New Jersey (SONJ) documentation of America during and after World War II. I have relied on Steven Plattner's history of this documentary project—the largest nongovernmental documentary project in the history of photography—in the following overview.[3]

In 1943 Standard Oil of New Jersey hired Roy Stryker to direct a photographic documentation of the uses of petroleum in America. Plattner quotes a Standard Oil official responsible for the project as asserting that the photographs were "to concentrate on 'the human part' of the company's work, particularly the everyday lives of its workers and their families . . . [to] foster the impression in the public mind that Standard Oil was composed of 'human beings like everybody else.' . . . Standard Oil sought to project a public image as a company that 'is a good citizen . . . that always works in the public interest.'"[4] George Freyermuth, Stryker's boss at Standard Oil, stated that "we [Standard Oil] wanted newspapers and magazines and others that wanted pictures to illustrate anything to come to us as a source for them and, over the course of time, to therefore identify Standard Oil of New Jersey as doing a whole range of things: taking care of little babies and promoting medical centers or being interested in what was going on in the playgrounds. We wanted them to know that we were people too and it had that effect."[5]

Stryker had spent the previous ten years directing photo documentation of the federal government's rural resettlement programs for the Farm Security Administration (FSA).[6] He directed the SONJ project largely as he had directed the FSA in the thirties. He hired skilled photographers, sent them to locations where he sensed there were good subjects, and loosely supervised their work.[7] He provided "shooting scripts" for the photographers, which were detailed directions to pho-

Fig. 2 Herbert Underwood, a farmer, pruning trees in his apple orchard, Wallingford, Vermont, February 1946. Photograph by Charlotte Brooks.

Howard Becker suggested several years ago that sociologists learn to analyze photographs by first concentrating on the act of looking itself. Rather than glancing and moving on, give yourself five minutes. Look beyond the quirky, almost humorous construction of the image. See that behind him are a small barn and a silo big enough for the corn of a ten-acre field. Who filled the silo and what animals will it feed? Behind the farmer are two more apple trees. These suggest the family's self-sufficiency; otherwise the farmer would not be spending a February afternoon preparing them for the spring. The farmer works alone on a winter day; his small herd of milking cows is mostly dried off and he has time for such chores. This reading is only the surface, a scratch at the meaning of the image, but even at this level the photo suggests the contours of a biography.

tograph industrial processes and their settings, including land, architecture, and social activities.[8] He insisted that his photographers study the regions and the industrial processes they photographed. He is quoted as saying: "They should know something about economics, history, political science, philosophy, and sociology. They have to be able to conduct research, gather and correlate factual information, and think things through." Then they can "go out and take pictures that mean something."[9] The SONJ photographers were encouraged to spend several weeks, if necessary, on a single project, supported on the then generous rate of $150 per week. Stryker felt that teaching the photographers some sociology and immersing them in the daily worlds they were to photograph would lead them to photograph taken-for-granted aspects of the company's operation. Subjects were to include the daily lives of the workers and the impact of the company on the towns or the regions. The irony is that, because oil was in one way or another everywhere, the photographers were able to photograph virtually anything of a social nature.

The SONJ photographers Plattner interviewed noted that Stryker did not appear to interfere with their interpretive and aesthetic freedom. That having been said, it appears that Stryker and the corporation had similar views of the project: the photos valorized the corporation and the way of life which it encouraged. Stryker may have been a left-leaning populist, but he was swept up with the promise of documenting a great story of the times.

Standard Oil's intention was not, however, simply to provide public information about the company. The photographic project, which eventually cost more than a million dollars, was "one small part of a massive public relations effort intended to improve the company's image."[10] This effort, initiated in the early years of World War II, was aimed at offsetting charges made in congressional hearings that Standard Oil was responsible for the severe shortage of synthetic rubber in the United States just as war production was increasing. Standard Oil, as it turned out, had signed a secret agreement in 1929 with the German firm I. G. Farben in which the company agreed not to develop synthetic rubber in exchange for the German company's willingness to forgo the American petroleum market. When this arrangement became public during the beginning of World War II, it created a public image crisis for the company. Senator Harry S. Truman is reported to have stated that Standard Oil's actions "approached treason."[11]

Standard Oil's decision to support what became a massive documentary study of American society emerged from these events. The company had completed a public survey which confirmed the dire state of its reputation, particularly among the "opinion leaders" of the society, including scholars, artists, professionals, and even business executives. Standard Oil first hired artists, including painters such as Thomas Hart Benton, to portray the company operations specifically related to the

Fig. 3 Farmhand sharpening ax, Route 25, Smithtown, New York, October 1945. Photograph by Gene Badger.

war effort. The fine arts approach was eventually abandoned and replaced with photography. The scope of the photo project grew to take on the general role of petroleum in American society. According to Plattner, the SONJ project was approved because key people in the Standard Oil public relations department were aware of how well the FSA photographers had communicated the success of governmental efforts at alleviating rural poverty during the depression.

Once these broad goals were agreed upon, SONJ essentially left Stryker on his own. Standard Oil's goals for the content of the archive seemed to have been met (indeed, the photos humanized the company's operations, and they show the far-reaching impact of petroleum), but not its goals for altered public opinion. Polls in the late 1940s showed that Standard Oil's reputation was essentially unchanged. Funding was drastically cut in the late forties and discontinued after 1953.

Even though Standard Oil did not feel it had achieved its purposes with the project, the photographs were widely distributed and undoubtedly influenced the popular culture of the day. They were made available for free to any interested party, with only the promise of a byline required for publication. There were essentially three distributions outlets. Photo stories using the images (Plattner cites thirteen) were published in a Standard Oil magazine called *The Lamp* in the 1940s. Second, the images were reproduced in nationally circulated magazines such as *Look* and *Life*. Finally, the images were used in high school and college textbooks. Plattner tells us that "In 1949 alone, nearly 50,000 prints were distributed by Standard Oil . . . [and were published] in nationally circulated magazines . . . [and by] 1949, project photographs had helped to illustrate more than one hundred books, especially school texts."[12]

That Standard Oil kept its hands off the project does not mean that the project could become a critical analysis of the company, corporate capitalism, or American society in general. The project remained an essentially upbeat report of work in its various configurations. The Standard Oil–sponsored images embody the simple verisimilitude often attributed to modernist documentary in style and content.[13] Ironically, this was Standard Oil's goal.

Stryker directed the project for seven years, until 1950. It continued for a few years after his departure on a smaller scale. The project generated more than sixty-seven thousand medium or large format black-and-white images, and more than one thousand color slides. Though profoundly important, the project has been studied only in Plattner's excellent survey, and the SONJ photos have been the basis of only four documentary reconstructions.[14]

Stryker hoped that the photographs produced under his guidance would eventually be regarded as a historically situated visual ethnography. Standard Oil operations became peripheral to much of the photographers' work, which probably led the company eventually to withdraw its support. For example, the photos I have collected for this volume have little, on the surface, to do with Standard Oil. They can, in fact, be read in part as the visual ethnography Stryker hoped to create, even as they carry and reinforce the ideology of the corporation. But largely for the same reasons the project inspired Stryker, the photos are useful today.

I spent a week in the Standard Oil archives at the University of Louisville, examining a sizable number of the sixty-seven thousand images to identify photographs of northeastern family dairy farming. When I came across these images, such as Sol Libsohn's photographs of the changing works crew eating dinner, I saw for the first time scenes farmers had described to me time and again. It was a remarkable experience. At the core of this project is my desire to share these photos, interpreted by those who had experienced their reality, and to use these interpretations to question the subsequent development of American agriculture.

Research with Photographs

I do not see the SONJ photos as having a simple correspondence with the past. Rather, I used the images to construct a history based on the memories of several people with quite different experiences and orientations.

That having been said, the research project benefited from several fortunate coincidences. The most important was that the SONJ documentary project was directed by an individual whose goal was to record normal events and occasions. In fact, inspired and informed by his friendship with the sociologists Robert and Helen Lynd, Stryker had long viewed documentary photography as "visual ethnography." The SONJ project was, in fact, born of his frustrations experienced as an economics professor in the 1920s with conventional, number-driven economics pedagogy, and his desire to find a way to communicate the abstractions of economics and the related social sciences through photographs.

The second happy coincidence was that the SONJ images portrayed agriculture in precisely the era in which I was interested: as it was moving from largely animal-powered and cooperatively organized harvests to the industrialized and noncooperative forms which are prevalent today.

The third coincidence which made this study possible was that the elderly farmers I knew in the North Country were the right age to have experienced the events portrayed in the SONJ images. They were then youngsters or young farmers, and many of their memories addressed how the generation before them had farmed. Thus, my study captures a moment which would otherwise go unrecorded. As Scott DeVeaux characterized his study of bebop, mine "lives at the shadowy juncture at which the lived experience . . . becomes transformed into cultural memory."[15] As I complete this book several years after I did many of the interviews, I realize that

Fig. 4 Putting up snow fences, Route 15, Steuben County, New York, November 1945. Photograph by Charlotte Brooks.

Fig. 5 The Acres men in about 1980. *Left to right,* Kevin, Mark (*top*), Carmen (*lower*), Trevor, Laurie, and Brian. Kevin currently owns and operates the farm. Carmen, Trevor, and Laurie are deceased. Photograph by the author.

most of the farmers who could relate their experiences to the SONJ images are either deceased or so elderly as to make the in-depth interviewing very difficult. In another five years, the generation I have written about will be gone.

Photo Elicitation

I assembled a history from the SONJ images using a method that has been called "photo elicitation." Photo elicitation, most simply, is a process of organizing interviews around photographs. The assumption behind photo elicitation, as in the case of a Rorschach test, is that looking at visual symbols may trigger meaningful memories in an individual. The meaning of the image, whether literal or imaginary, rests in the mind of the viewer.

Photo elicitation was first described by John Collier (who, as noted, worked briefly as an FSA photographer) in the first text on the use of still photography in anthropology, published in the 1960s.[16] Collier described using photographs to

Fig. 6 The Acres farm in the late 1980s, when they were milking about one hundred cows. The original barn burned in the late 1970s and was rebuilt as a free-stall, pictured here. To the lower right of the barn is a liquid manure pit, one of the first in the region. A barn for dry cows sits perpendicular to the silos, which contain chopped corn (silage) or, increasingly, chopped grass, called haylage. Each of these feeds ferments in the silo.

As farmers who have always optimized production, the Acres family uses use a "total mix ration" system, adding grains and protein supplements to homegrown feed to achieve optimal production from their cows. The ration-mixing system is tucked between the silos and the back of the barn. At the foreground end of the long, free-stall barn is the milking parlor. In general the layout of the farm reflects the simplicity of the factory farm, as described in the final chapter of this book.

To the upper left is a farm pond, which, for several months a year, becomes a hockey rink. Photograph by the author.

elicit information relevant to the social identity of residents, or the spatial organization of village houses. Collier assumed that different members of the same group had essentially similar interpretations of the same images. For the topics he studied with photo elicitation—for example, the social organization of a village–there probably was not a great deal of variation in what different informants would report.

In my study I have asked people with a range of backgrounds to look at photos of things they did a long time ago, to remember how they did these things, and,

Fig. 7 Jim and Emily Fisher, 1999. Photograph by the author.

more important, to remember how they felt about it. Their comments must be understood in the contexts of these backgrounds.

As I was working on this problem, I read Elliot Liebow's book on women who live in a homeless shelter.[17] Liebow worked with approximately the same number of people as I did, and he described his subjects' backgrounds so that the reader is able to judge how personal differences may have influenced what was said by different people. This is a simple problem with imposing implications. It is, in fact, the introduction of the logic of variable analysis to a qualitative study, in the sense that a reader is able to consider how speakers' background characteristics influence how they have interpreted the images. Of course, with a small number of informants, it is not appropriate to describe statistical relationships between independent and dependent variables. Still, it behooves us to see the knowledge and interpretations in the contexts of the personal histories and orientations of the speakers. Thus, I offer the following description of my subjects and how we worked together to produce this book.

Fig. 8 The Fisher homestead in the late 1980s, including the original stone house and barns. Soon after this photograph was taken the barn burned, nearly taking the house. Jim, who was born in the homestead, now lives with Emily in the ranch house on the right. Jim is barely visible behind the ranch house, tending to sheep which are about to the loaded into the farm truck also visible behind the house. In the foreground is their swimming pond, visited often by my family and many other neighbors. Photograph by the author.

My first interviews were histories of two farm families. I did these interviews, each of which lasted over several evenings, with Carmen and Marian Acres and Jim and Emily Fisher. These interviews helped me understand both the changing technologies of farming and how two families had adapted to these changes. It was in these interviews that I was introduced to the basics of farming: what crops typical farmers had grown for how many years, what machines they used, how they bred, milked, and slaughtered cows, how they had worked together in earlier eras, and what had happened in the meantime to their farms and the farms of their neighbors. These families were both going through major transitions when we did our first interviews.

Carmen and Marian had come to the States in the late 1940s from Canada. Carmen had grown up on a large farm and had spent some summers during his teenage

Fig. 9 Carmen and Buck Wagner, fall 1999. Photograph by the author.

years on combine crews in the Canadian west. His father had farmed with steam tractors in Canada in the early decades of the twentieth century, and Carmen adopted every new technology which presented itself. Carmen was the quintessential innovator. He was clever with machines and with investments, and he did not seem to care a whit for tradition. He bought machines when they were introduced, and he paid for them by working for his neighbors. Eventually he came to represent several farm machine companies and operated a sales and service business in addition to the farm. His large family (fig. 5) at that time was integrated into the farm operation; it was a happy and prosperous time, and by the mid-1970s Carmen was growing one hundred acres of corn (a tenfold increase in twenty years) and milking over a hundred cows. Several tragedies struck the Acres family before the farm passed to the next generation.

Carmen and Marian were not nostalgic about the past, but they felt that they were working longer hours on more and more expensive machinery for about the same profit they made in earlier years. Carmen, as the principal decision maker on

Fig. 10 The two-generation Wagner farm. The Wagners continue to use the original barn and silos, which still store much of the food needed for the seventy-eight cows they currently milk. The milking and manure systems are typical of what I have characterized as the "craft" mode. Photograph by the author.

the farm operation, did not envision an alternative. Above all, Carmen was a pragmatist. When he passed away as a middle-aged man in the early 1980s, many of us lost a friend and the neighborhood lost one of its most important farmers. His son, Kevin, continues the farm on the model established by his father. Kevin is now milking around three hundred cows and runs what is considered one of the most successful farms in the region (see fig. 6). Marian has moved across the road from the farm, where she runs a bed and breakfast while remaining active in the community and church.

Jim Fisher was the patriarch of a settlement farm with holdings, at one time, of several hundred acres. The Fisher land abutted our land to the northeast. It would be easy to write volumes about the Fisher family. In fact, Emily Fisher, who married into the family in the 1930s, wrote an extensive history of the family, which she graciously loaned me for my research.[18]

Their farm was established in 1815 by George Fisher, a Scottish immigrant. Jim's father, grandson of the immigrant, oversaw the farm from 1900 until 1936, when Jim took over. He had recently graduated (as had Emily) from St. Lawrence University, a private university in Canton, a nearby town. At that time, the Fisher farm was one

Fig. 11 Jim Carr outside his barn, 1993. Photograph by the author.

of the grandest in the neighborhood, established in manor-like stone houses, replete with horse-drawn buggies and one of the largest herds of Jersey cows in the county.

Jim and Emily's two sons farmed with them until they chose different careers as adults. As they approached retirement age, Jim and Emily were faced with either modernizing or scaling back, and, in 1977, they sold their herd. After that, they raised replacement heifers and sheep and gardened extensively, renting their land to neighbors. In 1989, their barn burned, ending the long history of farming in that branch of the Fisher family. They are now fully retired and remain active in their family and the community.

The Fishers represent the closest thing northern rural New York produced to an agriculturally based upper class. They are well educated and involved in local politics, their church, and the community. For many decades, their farm was one of the two or three showpieces in the North Country. Even now, with few, if any, agricultural activities remaining, the homestead is a symbol of the spectacular promise of agriculture there.

We did three long interviews together, one with the SONJ photos. Jim and Emily are ambivalent about the past, never glossing over the unrelenting physical

Fig. 12 Troy Mathie, Tom Lee's hired man, and Tom Lee, 1993. Photograph by the author.

work and appreciating how new machinery often made the most onerous jobs easier. But they also have a strong sense that as ecologically responsible farming has given way to factory farming, a rural community based on farming has died on the way. When I last visited them in the fall of 1999, I found them, at nearly ninety years of age, to be as keen of mind and delightful as they had always been.

Down the road a mile from where we lived, Buck Wagner and his son Carmen (fig.9) maintain the farm established by Buck's father early in the twentieth century. I saw Buck hundreds of times in the fields over the sixteen years I lived in the neighborhood, but I had never talked to him before we discussed the SONJ photos. He is a good example of a farmer who modernized moderately and runs a farm in what I have called the "craft mode." Such farmers continue to use many of the machines and techniques represented in the SONJ photos. When I last visited Buck and Carmen in 1999 they were milking seventy-eight cows, which is now a farm at the smaller end of the North Country scale. Their farming methods have not substantially changed in the past decades, and with a few supplementary income-producing activities (Buck, for example, trucks cattle) the farm produces enough to support two families.

Fig. 13 Willie fixes Molly's Rabbit, 1990. Photograph by the author.

Our first interview filled a long evening just before I left the North Country in 1990. We sat around his dining room table and the photos did what photo elicitation interviewers hope they will, which is to say that they stimulated memories and reflections about what it had meant to experience and participate in the worlds represented by the SONJ photos. The experience seemed similarly powerful for Buck, his wife Eunice, and their son Carmen. As had many of my neighbors, Buck had lived his entire life in the same house, tending the same land and milking cows in the same barns. Much has changed around him, but the essentials, for Buck and his family, have stayed the same.

I visited Eunice, Buck, and Carmen in the fall of 1999, to show them the manuscript and explain what I had intended to communicate with the book. Though we had not been close friends when I lived there, they made me feel as though I had become one in the meantime.

I knew Jim Carr (fig. 11) from nine years of Wednesday-night bowling. His farming experiences represent the other end of the agricultural spectrum. A generation younger than the Fishers and the Acres family, he grew up on the farm where he lived throughout his adult life. He was involved with changing works crews as a child and witnessed how the building of the St. Lawrence Seaway in the 1950s drew agricultural labor from small farms, hastening both industrialization and the demise of small operations. His own family stopped milking cows after this era, but they kept their farm machines (tractors, manure spreaders, mowers, and balers), and Jim continued to harvest several hundred acres of hay for sale to neighbors or for his heifer operation, which he continued through his adult life. He saw North Country farming from the vantage point of a man whose family did not modernize past the level depicted in the SONJ photos, but he retained one foot in the process throughout his adult life. Our interviews lasted through several afternoons. My guess is that he had not thought about the events and activities depicted in the SONJ photos a great deal before our interviews, but he delighted in the memories they brought. Just as this book was going to press, I learned that Jim had suddenly passed away, leaving behind many friends, including this one, and his family. This book, then, preserves his memories as it helps us understand how farmers using his methods interpreted the changes taking place during the 1950s.

Tom and Marie Lee, born in the 1950s a hundred miles to the southeast, had moved to the North Country during the 1960s. Tom Lee's mother, Dorothy Lee, wrote a lengthy history of a community called Pillar Point, where she and her hus-

Fig. 14 Arthur examines the manuscript, fall 1999. Photograph by the author.

band spent their lives.[19] Tom graciously lent me the history, which was based on diaries, census materials, and recollections. It was an extremely useful resource, though I did not interview Dorothy Lee.

Tom and Marie's farming methods were progressive yet based on values and methods of the past. They had a smaller herd but kept their cows longer than the more industrially organized farmers in the neighborhood. They grew grain and harvested it with thirty-year-old combines they bought at auction and restored. They did not push their herds for maximum herd average (the average amount of milk produced by each cow in a herd) but balanced their various inputs to achieve the most rational use of labor, land, and economic resources. Of the farmers I interviewed in the larger neighborhood sample, they appeared to be the most future looking, and yet they had the greatest understanding of the lessons of earlier farming.

Among the people I interviewed was Willie, an old friend I had made the subject of a biography in 1987.[20] Willie's father was a blacksmith, and Willie has spent his life fixing farmers' machines, in addition to farming, himself, on and off. He was used to photo elicitation interviews from our earlier research project, and the SONJ photos evoked memories of a familiar past. We did our first interviews with the SONJ photos just before I left the North Country in 1990, and when I returned several times to work on the book, Willie continued to help. He accompanied me on interviews with Art Henry and Mason O'Brien, who were Willie's old friends.

Our evening-long discussions were three-way give-and-takes, with Willie leading Mason and Art to more detailed reflections. Willie was a great research assistant. His interest in the past was based on a dislike for much of what had developed in the meantime, although, like the Fishers, he understood the harsh realities of earlier times.

Willie, Art, and Mason were born between 1916 and 1930. Often, their memories addressed the period before the mechanization pictured by the SONJ photos. Their experiences made them ideal spokespeople for the earlier eras. While none of these men were formally educated, they were knowledgeable and articulate research subjects. Art sold his herd only a couple of years ago, while Mason quit farming several years earlier. They too represent the other end of the continuum represented by the Acres and Fisher families: farmers marginalized by the advancing technology and increasing costs, who were eventually driven out of business. Their land had supplied land for their expanding neighbors.

When I last visited them in the fall of 1999, I found that age had been harsh to Mason; he sat in his living room hooked up to oxygen; he had smoked too many cigarettes for far too many decades. Willie, with whom I am in regular contact, remains as he always has, though the years, of course, have taken their toll. We ran the roads for two days looking for Art, now retired and living off the farm, hidden away in a neighboring town. These three old warriors represent a way of living that will soon be only a memory. Their contribution to this project has been invaluable.

In addition to these interviews, I also did shorter photo elicitation interviews with several farmers and, in particular, a most helpful informant, Rob Evans, who has sold farm equipment to all of my subjects as well as their neighbors. Rob's assistance extended over several years, on several visits and through many phone calls. It was Rob who knew the answer to what had happened to a particular farmer, or how a particular machine had evolved into another.

Visual Narratives

I based the photo elicitation interviews on photographs that showed how people worked. I was especially pleased to find photographs that recorded most of the steps of various farm jobs. I have presented these as visual narratives that record ethnographic detail, much as Mead and Bateson did in their study of Balinese culture in the 1940s.[21] I assembled visual narratives from photos made by different photographers, often working on different assignments and in different settings, so the narrative structure is entirely my own. Sol Libsohn and Charlotte Brooks did not see their photos in context with each other, as I have.

I have also included photographs which I did not use in interviews. These photos create a fuller sense of the historical era in which this study is situated. Many of these photographs would not have been useful in interviews precisely because of their generality, but for that reason they help the book move from topic to topic. I have thought of these images as similar to cutaways in a film, where a short film segment without a specific message is used to move a film from one topic to another.

In that spirit, I would encourage the reader to read the shared memories I have assembled while studying the photos, much as they might listen to spoken text as they observe the moving frames of an ethnographic film. Film communicates with emotional nuance as well as with explicit information. I would hope that this book can borrow from these filmic possibilities, as it moves between an analysis of, and a feeling for, the past.

One never knows when one has enough information for a project such as this, but I recall my mentor, Everett Hughes, admonishing me not to return for "one more summer" of fieldwork when I was studying hoboes for my Ph.D. dissertation. He said when you start hearing similar versions of essentially the same story it is time to stop. That was how I saw the farm interviews. I interviewed people from several backgrounds, and their contributions were lengthy and thoughtful. Together, these statements make a composite of a historical memory. Sociologists as important to the forming of qualitative methods as Herbert Blumer[22] reinforce the idea that a small number of well-informed informants are, in fact, a better sample than much larger samples of minimally involved subjects.

Guiding my research has been my assumption that social relations are influenced by material conditions. "Culture," thought of as shared plans and categories, emerges from people doing things together.[23] Farm cultures begin in family work; they are mediated through tools and technologies and directed to animals, plants, and the land. New tools and machines lead farmers to work differently, and they lead family groups to cooperate in work groups for specific jobs. These changes influence how people see their own involvement, the involvement of others, and the animals and land. As technology becomes more apparent, work groups become obsolete; some animals become obsolete; some forms of work disappear, and others emerge. The cultural definition of an animal and the land itself may be transformed. At a certain point, we should ask if the system which seems to have accidentally come into existence is logical or rational. This, indeed, is my purpose.

Historical Frameworks

To illuminate the industrialization which came to a head in the 1940s, I will describe structural changes in farming, including the impact of changing tools and machines, the evolution of crops and products, and the changing nature of agricultural labor. This includes women's involvement in and eventual exclusion from agricultural production. I emphasize the importance of agricultural technology in shaping how farmers worked, and how they lived together in families and neighborhoods, while acknowledging that culture, history, and other external events also influence which technologies will be adopted at any given time. The evolving structure of dairy farms is represented in table 1, which serves as a summary of much of the following.

Origins of the Family Dairy Farm: Pioneer Farms

Pioneer farms in the first years of the nineteenth century became, by the 1830s, what we now identify as family dairy farms. The pioneer farmers grew hay, grains, corn, vegetables, and fruit and bred and raised horses, oxen, cattle, pigs, and fowl, mostly feeding the farm family and the farm animals from their own produce. Eric Sloan's illustrated and annotated diary of an "early American boy"[24] shows how even a fifteen-year-old farm boy in 1805 was skilled at blacksmithing, carpentry, and various aspects of animal and crop work.

The family worked alone except for the hay harvest, which several farmers did together. Each would contribute a worker or two, and the informally organized crews harvested hay on one farm after another through a neighborhood. This was the first form of what came to be known as "changing works," which was marked by communal meals as well as shared working.

Fig. 15 Farmers looking at a planter, Flemington fair, Flemington, New Jersey, August–September 1951. Photograph by Todd Witlin.

TABLE 1 The Evolution of Northeastern Dairy Farms

	New England Frontier Settlement (1700–1840)	Early Mechanization (1850s–World War II)	Advanced Mechanization (1950–Present)
Crops and feed	Hay, grains (corn, oats), garden vegetables, fruit; farm is self-sufficient for animals and family	Hay, grains (corn, oats); some feed supplements are purchased for cattle; self-sufficiency diminishes	Corn (silage) and hay; some feed and all feed supplements are purchased; self-provisioning is rare
Animals	Cows, horses, oxen, pigs, sheep, fowl	Cows, horses, sheep, fowl	Cows
Products sold or consumed	Milk, cheese, butter, vegetables, honey, meat, fruit, juice, wool	Milk, cheese, butter, vegetables, meat	Milk
Method of harvesting hay	Cut with scythe in loose form, with help of neighbors	Cut with reaper in loose form by family	Mechanical baled or chopped for silo by family or hired help
Method of harvesting small grains (oats, sometimes wheat)	Cutting, stooking, and moving by hand; threshed by hand or horse	Cut with reaper; stooked by hand; moved to barn and threshed by neighbors	Cut and threshed by combine, which replaces all labor; then small grains are abandoned as crop
Method of harvesting corn (grain corn to silage)	As grain: picked and shucked by hand; sorted as whole ears or as grain; ground by miller	As silage: cut and bundled by reaper; moved and chopped for silo by neighbors	As silage: cut and chopped in the field with mechanical chopper
Method of housing and milking cows	Rudimentary barns; cows are milked in barnyard	Stanchion barn; bucket system for milking evolves to pipeline	Freestall barn; cows are milked in milking parlor
Manure storage and use	Not gathered or used	Raked from barn into mechanical spreader; spread daily	Liquid manure stored in pit; spread infrequently
Motive power	Horses and oxen	Horses and tractors; 50-h.p. limit	Tractors and self-propelled choppers; up to 200 h.p.
Labor	Family; neighbors help with hay	Family; neighbors harvest grain and corn and share other tasks	Family and hired labor; little informal cooperation
Farm size (acres)	80–120	80–120	80–1,000
Herd size (no. of cows)	15–20	15–30	15–1,000

The boy's diary shows that work was engaging and immediate in its impact. Along with his father and mother, the young boy helped feed and house their family. It is startling to read the maturity of the fifteen-year-old; by the end of the book he has proposed to a neighborhood girl, and he seems capable of establishing a new homestead!

Several studies suggest that the structural basis of the pioneer farms was less secure than is implied in the comments above. Since farmers did not rotate crops, farming quickly drained the land of its nutrients. Neither did they replenish their land with animal manure or other fertilizers. Cattle had little protection from the winter weather, and in the summer they foraged for their food on natural grasses in declining fields. In general, productivity was low, and minimal subsistence took a great deal of human labor.

Cheese Farms in Nineteenth-Century New York

The Erie Canal (built between 1817 and 1825) and early highways and railroads connected New York farmers to national and international markets and competition.[25] Immigrants from the British Isles brought cheese-making skills to their new farms throughout the northeast. In central New York, these were Welsh immigrants; our old neighbors in St. Lawrence County descended from Scottish immigrants who settled in northern New York between 1815 and 1820.

By the 1840s the regional agriculture balanced animals, crops, products, and family labor in a nearly ideal mosaic. At the core was the 115-acre farm (the average farm size in New York for nearly a hundred years!), which included between fifteen and twenty-five cows. Typical farms had two to five horses and perhaps a pair of oxen, eight to ten sheep, and six hogs. The farmers grew small grains (wheat, buckwheat, rye, oats, barley), corn, potatoes, peas, and root crops, such as rutabaga, which were used for supplemental cow feed. Their orchards included apples, pears, cherries, plums, grapes, and other fruits. Farmers raised poultry for eggs and meat, and bees pollinated plants and made honey. Large gardens were cultivated for home consumption.[26]

The typical farm of this era met the subsistence needs of the family. Farmers purchased only a few products, such as tea, salt and spices, and also had sufficient resources and labor to produce cheese for both home consumption and sale.

Immigrant farmers made cheese as they had in the countries from which they had emigrated. I refer to cheese making as a craft process because it required skill, judgment, and physical control.

Cheese makers judged the effect of different milks, and even feeds, on the taste and texture of the cheeses they made. The procedures for making cheese varied considerably, and often failed. The cheese makers "guesstimated" temperatures of heated milk and the solidity of curd and relied on practice to cut and press the curd to release whey but retain the desired amount of fat (there was no scientific measuring of butterfat until the Babcock butterfat tester was invented in 1890). Even the curing process was a matter of seasoned judgment and could produce either good cheeses or lumps of solidified, spoiled milk.

On pioneer farms women milked and took care of the cows and made the cheese. Their role was similar to that of the milkmaid on European farms. Making cheese in the preindustrial era was backbreaking: heating and pouring milk, repeatedly compressing curds, and continually turning the large, curing cheeses. The work involved heavy lifting and working with vats of boiling water. It required mastering a craft, and it took a heavy toll on women who spent twelve hours or more a day on the job.

McMurry notes that cheese making was defined as "women's work," and that it provided social and economic power for farm women. But she argues that the physical demands of the work were extreme. When cheese making left the home for the "crossroads cheese factories" in the 1860s, there was little outcry over the loss of economic and social power for farm women. At first, women shifted their work to the new cheese factories, but by 1900 women had nearly disappeared completely from the cheese-making labor forces. They found work off the farm, or they assumed other farm activities which did not produce the autonomy or social and economic power which had come at the heavy cost of home cheese making.

Farm-produced cheeses were made primarily for sale (often for international markets, even in the mid-1800s) and, to a lesser extent, for home consumption. The average farm studied by McMurry in central New York produced, in 1850, 4,800 pounds of cheese and 500 pounds of butter. Their cheeses sold for between five and seven cents a pound (for comparison, wages with board for a farm laborer in the mid 1800s ranged from nine dollars a month for a hired man on a winter season, to two and a half dollars a week for a dairymaid).

The cheese-producing farm of the mid-1800s was an ideal balance of resources, animals, crops, and labor in which "all activities circulated nutrients and energy endlessly."[27] The farm produced a rich variety of food for the family (I salivate at the description), including homegrown, pre-chemically-altered beef, pigs, and sheep, poultry and eggs, homegrown wheat for flour, vegetables from gardens, fruits from orchards, honey from local bees, and oats to feed the horses or oxen which powered the farm. The sheep provided wool for winter clothing.

But I believe the most important consideration is that the family cheese farms were estimated to produce as many calories as they expended, or even more, leaving out the contribution of human labor. The modern farm, by contrast, champion of machine and rationalization, requires between sixteen calories of energy to produce a calorie of energy of grain and seventy calories to produce a calorie from meat.[28] The earlier farmers used manure and local lime for fertilizer; they used their fields to grow the crops which became feed for their varied stock; they made a cash crop in cheese and used the leftover whey as additional animal food. Their labor was supplied by the family and neighbors, and animals powered the small machines which were used in planting and harvesting crops. The farms prospered economically and socially, building cohesive rural neighborhoods as well as successful individual family units.

Agricultural Technology

During this period, agricultural technology reached a second plateau. At the beginning of the era, farmers worked primarily with tools; by the end of the second stage, these tools were almost entirely replaced by machines. The modern era represents a third plateau of technological change, characterized by more complex and expensive machines and few, if any, tools.

This evolution is rather complicated, and the transitions from one plateau to another do not all happen together. For example, on pioneer farms, farmers cut hay and grain with scythes, as farmers had for several centuries. Family members raked and turned the hay until it was dry, then gathered it, placed it on horse-drawn wagons, and pitched it by hand into barns. The first hay mower, pulled by two horses, was introduced during the Civil War, coincidentally helping to relieve the labor shortage caused by the absence of men. The mechanical mower revolutionized the hay harvest. A skillful worker with a scythe could cut two to three acres of hay a day;[29] a farmer with a mower did twenty.[30] All subsequent handling of the hay (raking it for drying, picking it up in the field, moving it to the barn, and storing it in a loft), however, was still done by hand, sometimes aided by simple, horse-powered elevators. This system of hay harvest remained roughly the same until the era pictured in the SONJ photographs—nearly eighty years!

During the pioneer era, farmers also cut (reaped) grain by hand, using scythes, to which long fingers had been added to guide the cut stalks into windrows. Workers then gathered and tied the grain into bundles and stacked the bundles to dry. A worker with a fingered scythe (called a "cradler") and four or five bundlers could cut and bundle no more than three acres a day.[31]

Mechanical reapers were invented in the early nineteenth century and in use by the 1840s. The first versions of these wonderful machines, which McCormick sold in the 1840s for one hundred dollars, used two workers to cut ten acres of grain a day. Then, as in the past, the grain had to be gathered by hand, bound into sheaves, and stacked to dry. The efficiency of the mechanical reapers improved during the nineteenth century, during furious competition among hundreds of inventors and companies, but the single overriding breakthrough was the development, during the 1880s, of a mechanical reaper-binder. This horse-drawn machine cut the grain, gathered it into bundles, and tied bundles together with twine. Workers then only had to walk through the fields to place the sheaves of grain in teepee-shaped stacks to dry. This machine, like the mower described above, was in use with little change until the era portrayed in the SONJ photos, and it is still used by groups such as the Amish who do not use gasoline-powered machines. Again, we note a technological plateau of eighty or more years.

After it was dry, farmers had to move the grain to a central location, usually adjacent to a barn, and then remove it from the stalks. This is called "threshing," and in the pioneer farms of the early nineteenth century it was done the same way it had been done for centuries. Farmers sometimes beat the grain off the stalks with a flail, or let animals walk on the stalks to knock it off with their hooves. Some farmers used a thresher sledge—a coarse stone or wood plank pulled over the stalks, which ground off the grain. The grain then had to be separated from the chaff (winnowed) in another job, which, by today's standard, was extraordinarily inefficient. Workers threw the grain into the air so the wind would blow away the chaff, or threw it on the ground so that the heaviest material, the grain, would go the farthest, like the head of a comet.[32]

The first mechanical threshers were developed at the end of the eighteenth century in England. These machines worked on the same principal as modern threshers, using drums rotating inside a cylindrical grate to knock grain from stalks. Threshers are stationary; farmers bring grain to the machine and feed it in by hand. The first threshers were powered by humans turning cranks; then, by the beginning of the nineteenth century, they were powered by horses walking on treadmills hooked to gears on the machine. Eventually, horsepower was replaced by steam power, and, at the end of the nineteenth century, by gasoline engines. By the early decades of the nineteenth century, most grain was mechanically threshed, but threshing machines were incredibly expensive. In 1822 a horse-powered thresher cost five hundred dollars and only a few farmers could afford such a machine.

Mechanical grain threshing lasted until the mid-1950s. We shall examine it in detail, for it was well documented by the SONJ photographers. The remarkable fact is that the stationary thresher was another agricultural machine, along with the mower and reaper-binder, which operated little changed in basic design for nearly a hundred years and was largely replaced during the period documented by the SONJ photographers.

Thus, crop harvesting underwent several developments during the nineteenth century. This technological plateau exceeded the level of hand-powered tools. The second level was made up of machines, some of which were quite elementary and comparatively inexpensive (such as the horse-drawn hay binder), and others which were extremely complex and expensive, such as the thresher. This second level of technology preceded the tractor. The second-level technology was powered by horses, oxen, and even people, then by steam and by other forms of mechanical power.

Milking and housing cows also found a form during the mid–nineteenth century which was little changed for a hundred years. In the traditions which preceded American dairy farming (speaking here of centuries-old European peasant agriculture), cows, usually numbering no more than two or three per family, were housed on the first floor of two-story farm dwellings. The body heat from the farm animals heated the second story, where the farm family lived. In his novel *Independent People*, Halldór Laxness wrote a portrait of this arrangement, which persisted until the early twentieth century in Iceland.[33] On the farm Laxness described the cow competed with the sheep for first-floor space; she was cherished by the mother and the children but disdained by the father, who saw her milk as coddling a family otherwise able to subsist on dried fish and a little coffee. When the flock was starving before spring returned grass to the fields, the farmer butchered the starving cow so the sheep could live, and the family was devastated. In this moment, Laxness destroys any idealistic vision we might retain of the feudal farm.

The first farms of the New York region, however, were nearly as desperate. As the number of cows per farm increased, agricultural thinkers suggested fundamental changes in cow housing to improve the conditions of the herd. Consequently, barns were designed to hold cows for all the cold months of the year; the two-story structures typically included a mow in which loose hay for winter feeding was stored, and troughs in which manure was collected and removed from the barn. The first stanchion barns were designed in the 1830s.[34] In these barns, cows were confined to a specific place, either by a chain or by inserting their heads into parallel bars, which kept the cow standing or lying in the same place. While this sounds punishing, cows are easily habituated to such an arrangement. In addition, in the stanchion barn, cows are more easily milked. The milker moves from one animal to another and the stationary cows are less likely to kick the milker or the pail of milk. The stanchion barn requires a great deal of space per individual animal and considerable interior construction to adapt it to individual animals.

Later, I will describe how the modern era has largely seen the replacement of the stanchion barn with the free-stall barn and milking parlor. For now, I note once again that an important technological form, which came about in the middle of the nineteenth century, remains a viable technology. It limits the size of a herd which could fit into a barn most farmers could afford, but it also led to a pattern of animal-human interaction which nurtured the cow and created animal-smart farmers.

These are the main aspects of technological change leading to the great transition recorded in the SONJ photographs. Succeeding technological developments, such as the gradual introduction of the tractor, remain to be discussed. Nevertheless, I would like to highlight two aspects of this brief historical overview.

First, the evolution was from tools to ever more complex machines. This led to less human input per unit of productive output, but greater capital investment.

Second, all the technological changes described above (and thus to be described subsequently) influence how people work together. Some technological change led women out of agricultural production (as did factory cheese production), and other technologies altered how people worked among families (here I am thinking of the thresher). The horse-drawn mower eliminated communal hay harvests. The combine eliminated the communal harvest of grains. The eventual development of newer barns, manure systems, and corn-harvesting techniques further isolated the farmer from his or her neighbors. These themes will be developed in the analysis of the SONJ photographs and of a contemporary farm neighborhood.

Farm Size, Number of Farms, and Land Use

In her family memoir, Emily Fisher tells the story of her husband's ancestors' emigration in 1815 from Scotland to a farm in St. Lawrence County. Her husband's ancestors followed a social script that was well rehearsed by the farmers who had already arrived from Scotland. They cleared land and used the timber to construct log houses. Most immigrants started families, built better houses, and established self-sustaining farms within twenty years of their settlement. What is remarkable is that most of these farms stayed the same size for more than a hundred years.

Table 2 shows the number of farms, the size of farms, and the land in production in St. Lawrence County, New York. St. Lawrence County has been and remains one of the most important agricultural counties in New York, the leader among the fifty counties of New York in the number of farms and nearly the largest milk producer in the state. The data I review here represent the general pattern of northeastern agriculture: increasing farm size, decreasing number of farms, declining acreage in agricultural production, and increasing numbers of cows per farm.

Statewide, the average size of a farm in 1850 was 112 acres. A hundred years later it was essentially the same. The pattern in St. Lawrence County is a variation on that theme. In 1925 the average farm was larger than the state average, at 135 acres, about the size of the Fisher settlement in the mid–nineteenth century. By 1930 the average size had grown to 143 acres, but undoubtedly due to the impact of the depression, it declined slightly to 141 acres in 1935. At the end of the war, agricultural mechanization and consolidation began in earnest. By 1959 the average farm had grown to 210 acres. This figure had increased to 290 acres by 1990.

Since nearly all the potential farmland was in agricultural production in 1910 in St. Lawrence County, the increase in the average size of farms has been due to farms' absorbing their neighbors. Farm size, however, should be thought of as a calculus including total acres, the number of cows milked, the complexity of planting, harvesting, and barn technology, and even how much feed is bought instead of grown on the farm. Each of these elements pushes and pulls on the other. A farm may be in balance, as they were for over a hundred years, with fifteen to thirty cows and sufficient land to grow most of their food with the available technology and labor. As technology changes (either in agricultural machinery or barn type, for example), more cows may be added to produce the milk to pay for the change. Land may become available when a neighbor chooses not to or is unable to adopt the new technology. Farmers also often rent their neighbors' land, which is not reflected in census data. Consolidation of farms may thus be greater than reflected in these data.

The history of Jim and Emily's farm highlights the evolution from diversification and self-sufficiency to greater specialization, which eventually led them out of farming altogether. For many decades they had one of the leading farms in the region. The decisions that led the Fishers away from farming, however, were social as well as economic. As a family they may simply not have found the mechanized forms of farming sufficiently to their liking to continue. In addition, there were no sons or daughters who chose to continue as full-time dairy farmers, which obviously influenced their decision. Nevertheless, we see in the following history a charac-

TABLE 2 Changing Farms in St. Lawrence County, New York

	1910	1925	1940	1955	1959	1964	1969	1975	1990	1995
No. of farms	8,224	7,583	6,236	4,380	3,426	2,895	2,190	1,792	1,367	575
Average farm size (acres)	100	115	117	NA	210	230	248	259	290	NA
Land in production (acres)	1,061,516	1,012,449	991,642	NA	NA	664,649	543,494	NA	493,073	396,721
Cows per farm	12	12	14	14	19	22	32	35	46	54

Sources: U.S. Census of Agriculture (1925, 1945, 1954, 1959, 1969, 1992); Agricultural Census of St. Lawrence County, 1992; Huff (1996); New York Agricultural Statistics, 1996.
Note: NA, not applicable.

teristic evolution of the North Country dairy farm. It also demonstrates the extent to which a local agricultural economy was dependent on local economic enterprises, such as milk-processing plants or cheese plants, and local markets which may have come into existence because of health concerns and the perceived healthfulness of the food they were producing. Ironically, their specialty, the high-butterfat Jersey cow, was itself a victim of the cholesterol fears which developed during the 1960s. Thus, in the following history we see a microcosm of change and eventual decline on a typical and successful North Country dairy farm.

Jim: Dad farmed 1900 to 1936, which is when I came into the picture. The community was made up of smaller farms. Chipman had a butter factory which was the hub—the economic center of the community. All the farmers produced small amounts of milk, separated the cream, and the cream was delivered to a little plant in Chipman. That went on until about 1920. And then the Chipman plant was taken over by a large milk company—the Waddington Milk Company—which operated for a little while and then closed. Then the fluid milk went to either Madrid or Waddington to a larger plant. Waddington was a manufacturing plant. It condensed milk and shipped manufactured milk, and there was a cheese plant there too. Madrid was a fluid milk plant. Its milk was bottled; some went in bulk to New York.

My father milked from twenty to thirty cows. And they had a small flock of sheep. They had the maple sugar operation, as did many of the farmers. Bill Day's sugar bush was next to ours, and right across the creek was Myron Fisher, who had a small sugar bush. So that was a small income for everyone, but part of the picture. A number of them raised turkeys too, at that time. "Turkey day" in the fall, in the small towns, in Madrid and in other towns, was a big day. Everyone killed the turkeys, delivered them, and they were packed off to New York. These weren't large incomes, but incomes in general weren't that large then, and so the extras would be substantial. Dad used to say he hoped he could make enough with the syrup to pay the taxes. Even though they weren't as much then as they are now, taxes are always something you have to meet. . . .

It was all hand milking, until about 1924. My dad had one of the first milking machines. It was quite the same, really, as the modern ones except that you milked into a pail which you dumped into a tank. The herd size probably increased a little bit when we got the milking machine. It wasn't a dramatic increase. Our herd had gone up to thirty-two by then. We kept it about thirty-two until I came home in 1936; then we moved it up to fifty cows, which was one of the largest herds in the area. We did cost surveys in those years through extension, and there were not over thirty farms in the country larger than that, at that particular time. And that would hold until there was more mechanization, and then immediately the typical herd sizes went up to seventy-five, ninety, a hundred-fifty. . . .

Our next big increase was when we added on a extra long line of cows—seventeen—and that made it forty-nine stanchions. We had another older barn where we put dry cows, and did a few milkers by hand over there. I can remember milking fifty-five sometimes.

We always had help. We had usually two men—my father and I and two men were the farm crew for a fifty-cow herd. Usually there was a man with his family in Murray's house; that was our "help house." Then we usually had a single person who lived with us. And you'd wonder why we needed so much help, I'd guess!

We got our first tractor about three years after we were married; that would be about 1941. Up until then, the work was all done with horses. Three-horse teams were the common thing. Murray was two years old when we bought the old John Deere tractor—a new one, rather! It cost eleven hundred dollars for the tractor, the plow, and the disc harrows. Murray was so intrigued with the tractor! We started it by spinning a wheel on the side. We didn't realize it, but he watched us do that. It was parked practically in front of the garage and next morning, all of a sudden, we said "it sounds like the tractor is running!"—that two-cylinder clunkety-clunk. And here he was standing there—it was new and took right off when he spun the wheel!

In those years we had a cream market. We separated the cream from the milk here on the farm and kept it refrigerated. Three times a week it was taken to the railroad station in Madrid, transferred to insulated milk cars, and then shipped to Saranac Lake, where it was used as a good food for the tuberculosis patients who were housed in all those sanitariums. It was a special market and we were fortunate to have it. The cream market gave us a little edge over the regular milk markets in the area. But as soon as penicillin came in, Saranac Lake changed its face completely. People suddenly didn't need to go to Saranac Lake to be cured of tuberculosis. Not that they were really cured there—but at least it alleviated it. A number were cured!

As long as we were separating our milk, we had pigs, which we fed with the skimmed milk. That was a great bonus, because then you weren't shipping all of the product off the land. You were utilizing your skim milk, feeding it to hogs, and getting wonderful hog manure to put back on the farm. It grew great pigs; we had a great income from our hog operation. That was a lot of work too, a lot of hard work. Cleaning out all of those pens!

We had two big barrels where we put our skim milk, every day. They were next to the pigpens, and the milk would sour itself. After the milk was dipped out and mixed with the corn meal, it was fed to the pigs. There were always a few quarts of curd and sour milk left over; the new skim milk was added and by morning it was sour. It was then fed back to the pigs—it was an ongoing procedure.

The cash crops were the cream and the pork. We sometimes had ten brood sows. They bred twice a year with eight to ten pigs each, so that made quite a large number of pigs. We sold some when they were baby pigs and the balance would be fattened up on soured milk, butchered on the farm, and sold to the local meat market in Ogdensburg and Madrid.

Dad was never much of a gardener. He used to smile at people who wasted time in the garden because he thought that he had bigger and better things to do than working on a small plot. That was amusing to me because I liked to garden. Well, we had a garden but it was left up to the children, including me, and my mother. Dad took a dim view of the cash benefits—there was not much of a market for grown produce. It's changed now. But when he retired he really liked to garden—he became our gardener!

We grew all our own meat, and had a variety—pork, lamb, beef. We made maple syrup. We grew a large plot of potatoes and sweet corn—that sort of thing. We raised chickens for our own use, and they would pay for themselves with their eggs. And then we always had a fat hen for Sunday dinner.

After the sanitarium market closed, we were fortunate to get a Jersey market in Ogdensburg for a number of years. It was an all-Jersey market; our higher butterfat was valued. It was a whole-milk market, and that eliminated the skim milk and thus the hog operation. That was about 1938, because I can remember in those days we were producing about forty quarts of cream a day. We took the milk to Ogdensburg ourselves in a three-quarter-ton truck. We did that for years. Twelve to fifteen milk cans a day. It was all worth it because it was a fluid market. We would get probably a dollar a hundred more than the regular market. It was bottled and put right out onto the street the next day. It wasn't penalized as being a surplus milk, and put into manufactured products. That worked fine until homogenization came in, and paper cartons. Suddenly people couldn't see the cream lines anymore. And once you hide the cream so that a lower fat milk is just as salable, the milk factory could buy it much cheaper from farmers who used the Holstein cows. Then there was the cholesterol scare. The AMA began to recommend that people ought to have a low-fat diet. *Reader's Digest* came out with their damaging article in those years on the effect of animal fats in the body. The Jersey cow, of course, produced much more butterfat, so these new attitudes signaled the decline of the Jersey cow for awhile! That would be probably 1950.

When I was a child, there were more Jersey cows in America than any other breed. They are an efficient, small animal, and they're easy to handle, they have a nice disposition. And when we went into this other cycle of the homogenization, the low-fat milk—not as much paid for the butterfat differential—the Holsteins just flourished. And now, the Jerseys are about 10 percent of the population. The Holsteins came in and produced thousands of pounds of more milk with less food value compared to the Jerseys, and really created an overproduction. And that has caused a lot of these problems we have today!

We did get a bulk tank in 1960, but we really didn't increase the herd size at that time. In the late thirties we were at about fifty—we maybe did increase it to about sixty—that was the most we ever milked.

But that wasn't enough to afford the scale of technology we needed. We used one of the old wood silos, and we then put up one cement and stave silo in the early fifties. We didn't go out for silos for grass and corn. We tried cutting the grass before it was mature, wilting it slightly, and then field-chopping and blowing it into a silo. But it has a terrific odor—just awful. It is a pungent smell and it permeates your clothing, so that after you've seen it you really have to take a shower, otherwise you reek of grass silage. I can smell it on farms that feed it at night and I really don't like it. You would just live in a whole different atmosphere. Our silo is fairly close to the house, and for two or three weeks when we were putting up grass silage if the wind was in that direction we got awful silage smell. And there was quite a bit of runoff from it that was foul smelling. But the cattle liked it and milked well on it. Unless we would have the glass-lined Harvester silos, which enabled you to put it up greener and preserve it without that awful smell, you pay a price. We just decided to skip the grass silage stage altogether. We went back to corn silage and baled hay.

In 1960 we put the bulk tank in, and it lasted until 1977, when we sold the herd. To make it profitable we probably should have moved up to ninety or a hundred milkers, which would have meant a whole new level of technology—a whole new set of things.

If you are going to mechanize, and go the large route, you really need to have a partnership operation of some kind. One person can't really do it very well alone. You need to have a son or two that are interested or very good help in a managerial situation. When you look at the farms that are operating very well now, quite often it is still a family unit.

Actually, I can remember . . . I thought it would be great to have two sons because I'd been an only son and I wished that I had someone to help me and go on with the farm. So we were delighted when we had two sons. I can remember some of my happiest feelings would be when on a lovely day when we were haying, if I was mowing and I looked over and saw one son raking and the other would be baling—I would think "What more would one want! Here we are on a lovely setting; the family all working together." And there are a lot of advantages for those families who are enthusiastic about a farming venture. They are urging now that you should start training your children when they are very small to field part of the operation, so that they grow into it and have a feeling of belonging and a wanting to do it. And that was what we tried to do.

So we were kind of frustrated—here we were at late middle age being left with the operation of the farm, at a time when we had just made a big investment to go on, and to modernize. We had bought a bulk tank, the silos, and more field equipment. We had field choppers, but not huge tractors like they have now. We had tried to minimize our inventories of machines; we had bought secondhand tractors—not the forty thousand jobs that are around now. Murray was good at shopping for secondhand tractors that are reasonable in price and were adequate, really, for our sized operation. But as far as being able to go on and manage the equipment and the machines and the animals, we didn't want to, and decided that we would try this route. And actually it's been good—we've enjoyed being here and seeing that the farm has gone on, in a different way!

Carmen and Marian Acres, about ten years younger than Jim and Emily, approached technological change differently. It was in Carmen's temperament to innovate by adopting whatever new machines made themselves available, and his large family provided, for many years, workers to help in planting, harvests, milking, and other routine jobs. Thus Carmen and Marian's history can be read as a contrasting path through the numbers in table two.

Carmen: When we arrived in 1948, the farm consisted of a hundred ninety-eight and three-quarters acres; we were milking twenty-five cows. We had about twenty-five acres of oats, ten-fifteen acres of corn, and at that time we pastured all the time, so what land we didn't keep for hay the cows pastured.

That general picture started changing about nineteen probably fifty-three or four; somewhere in that time. Garnet Beckstedt bought a corn chopper, and he went around filling silos. . . .

Then we got our chopper and we increased the herd to thirty-five or forty. We just kept getting more feed from our land, adding on to the barn and putting more cows in it. When I started out at first I had only ten cows. But in July we bought ten more. I can remember saying, "If we get that side full, that will be it, that's all we'll ever milk." That was twenty-five.

Fig. 16 An abandoned farm. The house (*lower left*) is a burned shell with a metal roof; adjacent is a used house trailer housing nonworking rural poor. Across the road is the dilapidated barn used by the neighboring farmer to house dry cows. Cows and round bales of hay are in the upper areas of the photo. In this single image several historical narratives are born out. There was a point when the farmers gave up on their previously workable barn; that a farmer began to fix the roof with metal but never finished the expensive repair expresses the often desperate ambivalence of the small farmer. At this point, the whole farmstead should be bulldozed. Photograph by the author.

I bought the baler after I bought the chopper. At that time, we got ten cents a bale to do the neighbor's hay, most of the time. After I got done haying my own, I'd be filling everyone's barns—putting the baled hay on top of the mow—trying to pack in the last of it, you see. I used to do four or five hundred bales at the least—I more or less paid for the baler that way.

In forty-eight we milked with a bucket machine. The milk would go into a pail that we dumped into 10-gallon milk cans. We'd set it into a milk cooler. Then, in fifty-nine, we put the bulk tank in. Nineteen sixty-one, we put the milking parlor in. We went up to about forty-five milkers, I guess, and then we kept coming a little more until we were at about eighty-five, ninety. I guess we were up to pretty near a hundred when the barn burned.

We were increasing all the time. We'd built another row down the other side of the barn, so we were milking thirty-five. Because everybody thought it was funny why I'd put in a "couple-six" milking parlor to milk thirty-five cows. So then we kept increasing gradually. See, we put the milking parlor in and then we still were tying the cows and milking. I think for either two or three years before I turned it around to a free-stall barn. I think we were tying cows for about three years after that. It would be sixty-five or sometime after that we changed over to free-stalls.

About that time, when the bulk tanks came in, I bought the Joyce farm, and I farm the Baker place; really three farms in one. Plus, I had a field of Victor's for a little while, and next year I'll farm a field of Bill Day's. That one below Wayne's—I guess Bill wants me to take that. He's got it all plowed for me, he said the other day.

The events described in these two accounts show two versions of agricultural expansion, small parts of a fundamental historical transition. As shown in table 2, from 8,224 farms in 1910 there remained 575 in 1996, which represents a loss of 93 percent! The rate of farm loss in recent years has increased sharply, and the decline in the number of farms following World War II until about 1960, the focus of much of our attention, was extreme. But it is important to remember that each farm has a different focus and orientation brought by the farmers themselves, and we must not be content to interpret change simply through the numbers in a table.

Still, we recognize one overriding trend. The increase in farm size has not kept pace with the number of farms which have gone out of business. The result has been declining land in agricultural production. In 1910 there were well over a million acres in agricultural production in St. Lawrence County; in 1992 the acreage had declined to under four hundred thousand acres, a drop of more than 60 percent. In St. Lawrence County, an area of low population growth, mostly centered in the towns, the land released from agriculture has gone largely unused. Fields arduously cleared a hundred years ago have returned to brush and will soon again be forests. A few lots are sold for country houses, and more are used to park dilapidated house trailers. Behind these sad lots are the crumbling barns of the once working farms.

It is these structural changes which form the framework for my study. All the following stories of culture, technology, social relations, and ecology trace back to the numbers in the preceding tables and images and the outlines of the histories of the Fisher and the Acres families.

The Machine in the Garden

Fig. 17 *(previous pages)* Dairy farmer on horse-drawn sleigh waits for rotary plow to pass him. It is important for dairy farmers that roads be cleared, as milk must be delivered daily to the creameries. Lewis County, New York, January 1947. Photograph by Harold Corsini.

During the era photographed by the SONJ photographers, tractors largely replaced horses on the dairy farms of the northeast. The transition began earlier and it lasted longer, but the critical step in the evolution was captured by the lenses of Charlotte Brooks, Sol Libsohn, and the others.

My subjects talked about what the transition meant, and I will present their remarks as commentary to the images which follow. To provide the context for interpreting those remarks, however, a preliminary discussion of the wider implication of the transformation from animal to machine power is necessary.

Sociologists have long studied the transformation of rural ways of life to urban forms of association. These are, in fact, some of the most common themes of nineteenth-century sociology. Whether it be Tönnies's *Gemeinschaft* and *Gesellschaft*, Weber's concept of "traditional forms of authority" giving way to rational-bureaucratic forms, Marx's understanding of the evolution of feudal to capitalist society, or Durkheim's explanation of the change from mechanical to organic social solidarity, the essential argument has been the same: rural life, once characterized by long-term and face-to-face association and by customary ways of doing things, has largely given way to the anomic life of cities, strangers interacting through formal contracts, and masses crowded together but knowing no one. While not all people moved from rural to urban areas, of course, the character of urban life, to some degree, has permeated most of the society, including rural areas.[1]

A key part of the transition has been from knowledge of animals to knowledge of machines. Animals belonged to the world of tradition; the urban world was a world of machines and the factories that made them. The question of the aforementioned transition of forms of knowledge has not been seen as "what has been lost," partly because American rural sociologists have largely focused on traditional ways of doing things as problems to be overcome. There are several reasons for this, including the generally held assumption that the mechanization of agriculture represented progress. Since its inception, the discipline of rural sociology in the United States has had a formal relationship with the Department of Agriculture, which has taken as its mandate the extension of what is considered modern thinking into farming. The relationship between rural sociology and the U.S. Department of Agriculture has led to a discipline with few competing paradigms and methodological narrowness.[2] Thus, my intention is to examine the transition to mechanized agriculture not as an indication of increasing efficiency, but as a change that represents loss as well as gain. Some of that loss and gain is related to animal knowledge, the role of the horse on the farm, and the transition to tractors: what it cost, what it made possible, and how different farms either succeeded or did not succeed in this transition.

Knowing Horses

I draw upon symbolic interaction for this discussion. Symbolic interaction, briefly described, leads us to understand that people send out communications (words, gestures, expressions, and tones of voice, to name the most essential) and other humans pass these communications through a kind of interpretive grid we sometimes call "culture" to give them meaning. We react by assembling our own response, based on the interpretation of a message we have built. While there are libraries of studies of person-to-person interaction, the ideas of symbolic interaction have not, to my knowledge, been applied to the interaction between people and animals.[3]

Applying symbolic interaction to humans and animals is difficult because we can only imagine how the animal interprets our communication. We do believe that animals interpret our attempted communication, which is probably why we suspend our disbelief over the "animalness of animals" and devote ourselves to pets. We do not know how the animal interprets, but we imagine it.

The advantage humans have at imagining the animal as an "other" makes it possible for a comparatively weak human being to control a much stronger animal, such as a horse. The person who works with the horse anticipates its actions, but the person also influences the horse's actions with his or her own. Ronald Blythe, who studied villagers in England, described it this way:

> The horses were friends and loved like men. Some men would do more for a horse than they would for a wife. The ploughmen talked softly to their teams all day long and you could see the horses listening. . . .

> The horsemen . . . made the horses obey with a sniff from a rag which they kept in their pockets. Caraway seeds had something to do with it, I believe, although others say different.[4]

In the family farms of the North Country, farmers typically had a team of two or four horses. They worked these animals for at least twenty, and as long as thirty, years. They were bred from the mare of a previous generation, or purchased for thirty to one hundred dollars. The foal entered the family like a new baby and was likely the pet of the farmer's children. The horses were trained to the bit, the harness, and the discipline of work; the farmer had to bend the will of the horse to his own. This meant a kind of intimate knowledge, the ability to read the mood of an animal and to adjust to it.

The farmer built the work of the day around the strength and stamina of the horse. In other words, he understood the work partly from the horse's perspective, for what is hard for a horse is different from what is hard for a person. The ability to see through the horse's eyes is a function of experience and sensitivity. Like many forms of mastery, this farmer's knowledge added to his identity and, likely, to the pleasure of his work.

It even went further than that. Horses sometimes are thought to have a "second sense" that guides a higher level of understanding. Buck Wagner told this story about how horses avoided trampling him to death when he was a child:

> When I was a kid I got run over by a team of horses. I was standing out there by the barn door and Bill Shearof was our hired man. At noon you unhook the horses off the wagon and put them in the barn—give them some oats and water and so on. So when he went to drive them in the barn he picked up a corncob to throw at somebody else. But I was standing beside the barn door—I was three years old but I can remember when he put the lines back in his hands he crossed them. He got the left line in the right hand. Well the horses weren't going just were he wanted them to, so Billy was kind of a rough little fellow—he gives that right line a yank and it was the wrong line, and it pulled the team right over on top of me. But I remember I was right underneath them and those horses were 'a tramping, and my Uncle Ralph, he run over and reached in there and grabbed me by the collar and pulled me out—and like I said I was just three years old. And them horses never touched me—and it was often said that a horse wouldn't. They done well not to step on me!

The working relationship between the farmer and the horse might last through most of the years of a farmer's working life. Just like one gains pleasure from working with another person, I am sure that the farmer had a similar connection to his horse. This is reflected in comments by Mary and Virgil about the team they had for much of the history of their farm:

> *Virgil:* We kept the horses after we purchased a tractor, both my dad and me. One of them lived to be thirty years old!
>
> *Mary:* Virgil wouldn't get rid of them—they'd served him and he wasn't going to get rid of them. He wouldn't give them up for fox meat, you know. Somebody would say: "Why don't you sell them for fox meat?" But Virgil said -they'd served him and he wasn't going to do that. He buried them.
>
> *Virgil:* I bought them from John Seamorst; they were second-year colts. He'd go out west and buy a carload. I'd say they were wild horses!—
>
> *Mary:* —They were wild!
>
> *Virgil:* I think he charged me thirty-five dollars apiece for them.
>
> *Mary:* There was a red roan and a blue roan. They were pretty! But were they wild! You couldn't trust them!
>
> *Virgil:* I had to put them into a box stall. They wouldn't drink. You had to set a pail of water in there and go in to eat and hope they'd drink!
>
> *Mary:* They'd run away . . . took the wagon way off over here, across the road . . . they didn't stop until they got stuck! Oh dear . . . we still hung right to it; kept at them. We had only those two horses, the whole time we farmed. They were pretty, though . . . they were so pretty.

The connection between horse and farmer has memory for both parties. This memory takes the form of the "personality," a pattern of behavior toward each other. As Virgil and Mary point out, this relationship was not totally predictable; the horses maintained some independence. Thus, we speak of the relationship as negotiated, like all those between people. When the horses were gone from the farm, part of what made the farm was gone with it. A kind of knowledge and skill became obsolete. A good farmer did not need to know how to use animals for power; it was machines and their maintenance that mandated attention and intelligence.

Horse power, for better or worse, limited farming to a small scale. In the West, where fields are flat and miles long, there is a logic to faster and more powerful ways to farm. Before tractors, farmers harnessed forty or more horses together to pull plows, combines and other large machines. It was an incredible achievement even to arrange the harnesses for this forty-horse team, and certainly not a simple thing to keep the animals organized.

Pulling implements such as manure spreaders or reapers with horses was also difficult because the mechanisms inside the machine were driven by the wheels, like the blades of the hand-pushed lawnmowers a few of us remember from our youth. As a result, these machines needed to be small and, from the perspective of a tractor-powered farmer, inefficient.

Tractors

Tractors replaced horses because they could pull bigger machinery faster. Thus, farmers could plow bigger fields with the same number of workers; eventually, they could use tractors to power machinery. But the history of tractors and their adoption is rather complex.

The first tractors were designed around the time of the Civil War.[5] These were huge, slow, and expensive steam-powered machines, particularly suited for the miles-long fields of the Canadian Midwest. By the turn of the century, these "steam plows" weighed twenty-five tons and could turn furrows across a thirty-foot-wide swath, taking the place of forty-horse teams. They were designed for areas where farming was more productive than the available workers and technology could service. The rock-free and fertile fields of the American and Canadian Midwest stretched beyond where the eye could see. Thus, the steam-driven tractors were common in Canada and in the States in the early stages of mechanization. Carmen Acres's father used steam tractors when he was growing up on a large farm in Canada:

> We had a tractor as long as I can remember. I remember the old International Titan—a big drum in the front with the steam coming out—the old steel-wheeled ones. They used to pull three plows. My father, digging out stones with the tractor, would be, hell, as far as from here to James's farm, because they had that thing on the front that would steer it down the furrow. They only used it for plowing.

Steam tractors, however, were too large, expensive, and awkward to be used on the family farms of the northeast. It took small and efficient gasoline-powered tractors to mechanize North Country farms. But the invention and development of the gas-powered small tractor was itself a long process.

Gas-powered tractors are powered by the internal combustion engine, invented in 1876. Internal combustion engines were both more efficient and safer than the steam engines common at that time. Steam power requires an external boiler, a worker to keep the firebox fueled and stoked, and a steady supply of water. Because steam engines give off sparks, they can ignite a dry field, and, because they depend on high combustion in a single chamber, they can explode. There were, indeed, reasons to develop an alternative.

The first gasoline-powered tractor was designed at the end of the nineteenth century; within ten years they were competing with steam tractors in public demonstrations. The first gasoline-powered tractors resembled existing steam tractors in size and power.

Henry Ford observed competitions between steam and gasoline tractors held in Winnipeg in 1910. Rather than focusing on whether steam or gasoline power was superior, however, he apparently identified the relative size of existing and potential tractors as most important. The gasoline tractors introduced by the Rumley Company at that time weighed more than twelve tons, moved at less than two miles per hour and cost over three thousand dollars. These enormous machines got stuck in muddy fields, could only maneuver in huge fields, and were far too expensive to be considered by most farmers.

Ford's Model T car, which cost nine hundred dollars when introduced in 1908, dropped in price to under four hundred dollars after Ford developed the world's first assembly line a few years later. It was the assembly-line efficiency and design simplicity of the Model T that Ford hoped to apply to the tractor.

During this era, few farmers owned tractors, but they began to adapt Model T's to farm use, using them for pulling plows, planters, mowers and reapers. Farmers also jacked these cars up and drew power from the rear wheels to drive threshing machines and other stationary machines on the farm. The adapted Model T was, in fact, the first small tractor.

Ford's first actual tractor (the Fordson) was introduced in 1918. It was an extraordinarily simple machine, with engine, transmission, and axle components bolted together to constitute an external frame. The first Fordson looked essentially like the same tractor purchased by North Country farmers in 1949, shown in several SONJ images. It was powered by a 20-horsepower engine, and cost $750. In the middle of the 1920s, during a severe agricultural crisis, Ford dropped the price to $395. This was the first mass-produced and mass-consumed tractor, but it was not universally adopted for technical and social reasons.

The Fordson represented a breakthrough because of its size and power. But it only pulled implements; it could not transform its energy to spin implements and drive other farm machines. The first tractor to have this capability was the International Harvester, which included the first power takeoff (PTO), a shaft geared to the engine of the tractor which rotates under its power. This feature had an extraordinary impact on farm machinery. Functions such as movable floors on manure spreaders, which had previously been driven by the rotation of the implement's wheels, could now be powered by the tractor as it was driven down the field or even as it remained stationary. The PTO unit and other improvements were introduced by International Harvester in 1924 in a tractor called the Farmall. Updated Farmalls were sold in the North Country as late as the 1950s. A few are still in use.

But even the PTO did not lead to universal adoption of the tractor on small dairy farms in the northeast. These inventions were developed during eras of severe economic crisis in American agriculture: the loss of European markets after World War I and then, beginning in the early 1930s, the American depression. Thus, most family dairy farms could not mechanize until after World War II. In Canine's words, "The sluggish pace of farm-equipment sales briefly increased in the late twenties, then plunged into an abyss. At Deere and Company, for example, total sales went from . . . $76 million in 1929 to . . . 8.7 million in 1932."[6]

The United States' entry into World War II deeply influenced the shift from horses to tractors. As had the Civil War, World War II pulled millions of men off farms. When their sons and hired men disappeared, farmers had two choices: to work more efficiently through machines or to go out of business. My subjects recalled that for those men fortunate enough to be available for farm labor during the war, wages increased from one dollar a day to one dollar an hour.

These events took place in a culture in which many farmers resisted mechanization for social reasons. They knew horses and they did not know tractors. The horse was a natural part of farm practices that had existed for at least a hundred years, perhaps for three generations of farmers. The farms were organized not only around the horse as power, but around growing horsefeed as part of the normal cropping practices. To abandon these traditions was to redefine farming itself.

The transition also represented an unprecedented financial commitment: the beginning of debt as a part of operating expenses. At that time, horses could be purchased for fifty dollars; by the end of World War II a 25-horsepower tractor with basic implements cost over one thousand dollars. This, in fact, equaled the cost of a sizable herd of milking cows and the horse-powered implements of the day. Buck Wagner recalled:

> I can tell you something else that you probably won't believe here but I've got the papers up in the vault in the bank. When my father came here in 1929 he bought thirty-eight head of cattle, cows and heifers; three work horses; and five pieces of horse-drawn equipment for $890. In 1929. And interest on that was—this is what I couldn't understand—6 percent in 1929. And it said on the paper that all he had to make was the interest twice a year. Back during the depression for three or four years you would see that all he paid twice a year was the interest. And then you would see where he paid maybe $40 on the principal plus the interest on $890. I don't know why, but one year he paid a lot.

Eventually, however, cultural resistance was overcome; the gain in efficiency made possible by the tractor brought the use of the horse to an end on all but those

Fig. 18 Farmers inspecting Ford tractor, New Jersey state fair, Trenton, New Jersey, September 1947. Photograph by Sol Libsohn.

farms which operate, as a matter of principle or religion, without internal combustion engines.

James Morrissey, a retired farm equipment salesman, provides another view of this transition. His memoir[7] describes selling tractors to horse-powered farms, which he saw as carrying progress to a stubborn and resistant culture.

Indeed, this view was reflected by several farmers:

> *Virgil:* I didn't like plowing with the sulky plow, and I couldn't plow with a hand plow. With the [horse-drawn] sulky, I wasn't too straight, either. My dad was better; he could use a hand plow. I never got the knack of a hand plow. In the dirt, below the dirt, everything else!

And for many farmers, tractors and horses overlapped:

> *Jim Fisher:* After we got our first tractor in '42 we still used horses. Oh, yes, a little bit. Because people still had the horses. They didn't mechanize completely right away. The team was still kept for drawing out manure—oh, you

couldn't get on without horses on the farm! People actually had horses a long time after they stopped using them. Because, as you say, there were also emotional attachments. The tractors and the horses probably overlapped for fifteen years.

A lot of the same operations done with the tractor were done with horses. And that worked well, too, because it actually saved a person. One man got on the wagon and wrapped your lines around the standard after you started, and got your horses headed straight—you could just chirp at the horses and they would move up a step and "Whoa!" and they would stop . . . you could work the field yourself.

Mason O'Brien made the same point:

Oh, yeah. We always kept the horses. We just used the tractor in the spring's work. Working up the land. Everything else was done with the horses. All the drillin'. And all the hayin' and everything else was done by the horses. Didn't cost nothin' to feed 'em! We grew our own feed. The tractor you had to buy gas for.

For others, the end of horse farming was an unemotional and rather inconsequential change. I asked Carmen Acres about the transition:

Doug: Did people hold on to their horses for a while after they didn't use them anymore?

Carmen: A lot of people kept a team for a long time, after they didn't use them very much. Even my own—I think they ran in the field for two years and I don't think I used them but once in a while in the woods, a little bit. I guess the last year I had them that's all they did do, pull out a few logs. I don't remember what I did with them! Or who bought them, or . . . I don't remember what I did do with them!

Doug: Maybe they just walked away one day!

Carmen: Oh, I must have sold them to somebody, or I probably traded them off on something. I know, I traded one off a wagon one time, that I brought from Canada—a driving horse. I traded that off on a rubber-tired hay wagon. I don't remember what I done with the others.

Whatever the attitude of the farmer, and whatever the impact the replacement of the horse with the tractor had on efficiency, with the end of horse farming, associated workers became as obsolete as the animals they serviced. The SONJ photographers photographed two of these, the blacksmith and the harness

Fig. 19 A tractor pulling another tractor, which pulls a corn chopper and a wagon—400 horsepower rutting a field on a wet fall, trying to get the corn harvested, October 1989. Photograph by the author.

maker. I discussed the work and social roles of these workers with Willie, who lived through these transitions himself. This discussion follows as the text for these photos.

Finally, it is important to see this transition in terms of ecological issues which surround them. At certain historical moments, farmers understand ecological relationships between land and animals. This is not an automatic connection; many farmers in all historical eras and in all settings have used methods which are not sustainable. Farmers gradually accumulate knowledge about the ecological impact of their actions and have sufficient surplus to afford to invest in practices with longer-term payoffs. According to McMurry, the middle to the end of the eighteenth century was such a time. Soil and animals were cared for in terms of their ability to recreate themselves, and, interestingly enough, the caloric input to the farm roughly equaled its output.

The transition to mechanized farming began a shift away from seeing farming as part of a natural cycle. Larger tractors do not simply pull heavier equipment

Fig. 20 One-hundred-twenty-horsepower tractor pulling disc, 1983. Photograph by the author.

faster. As the tractor increases in weight it is also more capable of compacting and rutting the soil.

Figure 19, which I made in the 1980s, shows a tractor pulling another tractor, which is pulling a chopper and wagon. It was a wet fall, and farmers, desperate to harvest their corn, were using the most powerful machinery they owned or could borrow. The deep rutting of the soil which resulted can be seen even from the airplane from which I made this photo. What cannot be seen is the compacting of the soil caused by the huge machines. The rutting mixes the living topsoil with the dead gravel which lies beneath it. The compacting of the soil keeps water from draining efficiently into the ground to feed the various organisms which live there.

Horses are not heavy enough to do serious damage to the soil, and because a team of two horses pulls small equipment, it can be maneuvered in the small and irregular fields of areas like New York's North Country. Willie often said, during the wet falls, that he could get any corn crop off any field in the North Country with a team and small equipment. Of course, the pace would be slower, but he would be making more progress than would the farmers stranded in the muck with their mechanical behemoths.

In that the Amish continue to farm without tractors, they present an interesting case study. Hostetler offers the following: "When the Amish were urged by a US government spokesman following World War II to use tractors to produce 'more grain for starving peoples of the world,' the Amish bishops issued a firm response. They asserted that they were growing more food with horse power than their neighbors were with tractors. Every member of their group had already plowed every square inch of his farm (by April), they said, while the neighbors with tractors were just starting to plow."[8]

When tractors were first introduced on American farms, the Amish rejected them with the comments "They don't make manure" and "They ruin the land." Most outsiders accepted the comments as harmless rationalizations, and yet no in-depth study has ever been made of the conservation factor in Amish horse-powered farming as contrasted to tractor farming. Even with horse power, the Amish with few exceptions are first to have their fields plowed and seeded. Tractors, the Amish say, compact the land, which results in reduced yields. Amish farmers who have bought land from the non-Amish have noted that the soil begins to work more easily after the third year. The land also begins to drain better, so it is ready for plowing earlier in the spring. Plant roots penetrate the soil better and crops survive better during periods of drought where the soil structure has not been destroyed by compacting.[9]

Finally, as noted earlier, larger tractors represent both a greater "caloric investment," as represented in the energy which was required to produce the machine, and an increased "caloric expenditure" in the energy required to power it. Thus, increasing the size of equipment makes agriculture less ecologically rational: more energy is consumed generating the food than is produced in the food. This irrational system can exist because food and energy costs are not priced in terms of their caloric values.

This discussion perhaps seems innocent in our current world, only because we are blind to the underlying natural costs of our farming. Economic irrationalities have led us to milk cows thousands of miles from where their food is grown and to treat their manure as hazardous waste rather than as replenishment for fields. Our innocence, however, leads only to a temporary bliss; an agricultural system based on vastly more caloric investment than it produces will eventually crash. We examine the transition from horse to tractor in part because it was the beginning of these profound changes.

The SONJ photographs are now presented with excerpts of interview texts, organized around the normal year's work. In selecting the interview segments I systematically compared all things said about a given photo to find the most representative and illustrative comments, having at my disposal over seven hundred pages of interview text. I have often included several speakers, to balance the nuances of interpretations. But, in the end, I have been limited by the space defined by the visual narrative built around the SONJ images. Like a filmmaker editing a film, I have been able to speak only while an image has made that space available.

ROUND BARN FARM.
EST.
1886
J.C.YOUNG

Horses and Tractors

Willie: I think one of the biggest reasons they started givin' up on the horses was they didn't want to take the time to take care of 'em.

Mason: That's right. I think that's the reason. Then they could keep three or four cows in the barn or they could keep a team of horses so they figured the cows would pay better than the horses.

Buck: I used to rake hay—I used to drive the hay-loader we used to have when I was six or seven, not much older than that.

I remember we had an English Ford tractor like that one (fig. 18)—it had running boards—and I can remember my father would be out cultivating and I would sleep down on this running board between the two wheels. There was a running board there that you could either stand on or step up on to get on the tractor.

Doug: That would have been in the early forties?

Buck: Yes, early forties. There is an old, blue English Ford. I can remember riding beside the livery rake, sitting there going to sleep on it. My mother would be raking and I would be sitting up there in that seat—you would sit there and you wouldn't be very old and you would go right to sleep there on the rake. And she would holler at you—to wake you up so you wouldn't fall off—of course in them days kids done more. She used to tell about when I was a baby—she'd be out stooking oats and put me in the fence corner with a bottle and I would sleep in the fence corner while she was stooking the oats.

Arthur: Well, then, you see, this is plowing. He does a good job too. Every one of them looks the same. A good pair of horses; you can do an acre and a

Fig. 21 Farmer and Belgian stallion, "Major," Greene, New York, November 1945. Photograph by Charlotte Brooks.

Fig. 22 Farmer plowing near Route no. 1, Allenton, Rhode Island, September 1947. Photograph by Gordon Parks.

half a day. Maybe a little more, dependin' on how good the ground is. How many stones are in it. I have plowed two acres a day. But then I had a good start; I had a better team than these fellows, maybe.

Stones. . . . What happens when you hit 'em? Well, up in the air come them handles . . . up in the air! They might hit your chest . . . your chin. Got to go slow and then back up and pull it back and start over again! That's good.

He's plowing land which is sod; it was a meadow the year before.

Willie: See where he is holding the line? He don't hold them in his hands. He ties them so they come around his waist when he is plowing. This unit across here is a "single tree" and the one that goes across that hook—the two single trees together—is a "whipple tree."

You move with the plow. When you hit a stone it picks you right off the ground—I was small and it picked *me* right off the ground. Your plow is set so it will go into the ground—there is a little wheel which keeps it from going too deep. A leveler wheel called a colter. You set this for your depth; if your plow was going too deep, you set that wheel down. It is not a big effort to do that.

Jim Carr: If you want nice plowin' you get somebody that can plow with a hand plow. They'll lay that baby right over and it'll be just as true as can be. You're going the same speed—you're not going fast. And she'll just roll her right over. When you look across the field all your furrows are just right.

Fig. 23 Ralph Smoker plowing with tractor, between Coatesville and Lancaster, Lancaster County, Pennsylvania, April 1946. Photograph by Gordon Parks.

Mason: Yeah, they are nice horses (fig. 24). I'll tell you they pulled hard! Team has to walk right along to keep it cutting.

Fig. 24 Paul Blake, a hired man on the S. B. Moore farm, at Skowhegan, Maine, cuts a field of clover, August 1944. Photograph by Gordon Parks.

Willie: If they slow down it don't cut as good. . . . You got to have a good walkin' team.

Mason: You gotta' have a team that'll strike right out and really walk!

Doug: How many hours would you use a team before you'd give 'em a rest?

Mason: Oh, you'd probably cut an acre and then hook off of that, and hook on to the side raker. Rake it up with the same team and that's a lot easier, of course. . . .

Willie: When you was rakin', that's more or less restin' your team. They's just walkin' then.

Mason: Yeah. And then when you got it raked up, when it got ready to put

Fig. 25 Horses and wagon in barnyard entrance of a farm near Lowville, January 1947. Photograph by Harold Corsini.

it in, you hooked on to the wagon and hooked a hay-loader behind and went to drawin' hay. We had two teams, see. We had one be mowin' and rakin' and then we'd have one drawin'.

It's hard to mow after dew falls. It don't mow good at all. Usually you'd have to wait 'til it dried out to mow. Even in the morning, if the dew was on, you had to wait 'til after the dew was off before you could do anything. Turn it . . . or rake, or mow or anything else, probably ten o'clock, when you started.

Willie: A lot of it depended on how much undergrass you had. If you had a lot of undergrass you couldn't mow with either one of those.

Mason: Yeah, the first part of the hayin' mowed pretty good but then towards the end, you'd get like a second cuttin' in the bottom of it. It would get thick . . . and Jesus, some of that stuff was hard to mow!

Willie: Just like runnin' into a cement wall.

Mason: You bet!

Doug: So if you got five hours out of these horses in a day, that'd be it?

Mason: Oh, they worked longer than that! We would draw hay after milkin'. Draw in hay 'til the dew started fallin'. You can tell when the dew starts fallin'; we'd just look at the wheel on the hay-loader or the wagon and if it turns dark colored . . . it's startin' to get damp. And then you better quit.

Willie: A good team would outwork two men in a day. They could outwalk you any time!

Mason: Oh, we had two good teams! Percheron. Good teams. We had one team we always mowed with; they'd walk about six–seven mile an hour. And mister, the old mower'd just rattle; never slow up.

Willie: Hit a woodchuck hole, bounce him just 'bout off the seat.

Mason: The only bad part of that was, if you was plowin', they wanted to go the same way with a walkin' plow. We always had a job to stay behind 'em. I never liked to plow with 'em, but I did like to mow with 'em.

The horses had the hard part of that job. Na, it wasn't that much work to it. That or rakin'. Drawin' it was the biggest job in hayin' . . . gettin' it into the barn. And gettin' the mow to lay.

The daily schedule was built around the horse. The work was balanced between what could be accomplished with the modest power of the horse, the number of animals a farm supported, and the amount of land a farmer planted and harvested. The following passages show how even spreading manure was a routine which depended on knowledge and skill.

Emily: Looks like the old times (fig. 25). Spreader with the steel wheels.

Jim: Yes, that is a steel-wheel manure spreader.

Doug: Did you spread manure every day in the winter?

Jim: Oh yes.

Emily: Every day they could.

Jim: You would get up, go to the barn, and if you had a helper, usually one person would start to milk; the other would feed the silage and grain the cows. Then you would milk the cows; put the milk in the cans—strain it through cloth filters, put it in the cans, and have it ready to go to the factory. The night's milk was refrigerated. Way back, in the early forties, before everyone had electricity, there was an icehouse near the barn where ice was stored—packed in sawdust. After the milk was placed in the water, there would be a trip to the icehouse, and a big cake of ice hauled out from the sawdust and taken in and put in the milk tank. The ice had been put up the year before in the winter.

Then you went in for breakfast, and right after breakfast you would come out and hitch up the team onto the manure spreader, drive through the barn, pitch the manure on from each side. Then if the snow wasn't too deep, the team went right out onto the meadows, wherever you wanted, and it was spread all winter long, until the snow got too deep.

If it was quite deep you would go out and make a trial run and mark out the

Fig. 26 Carroll Lang and farmhand are sawing up firewood, Lang's Gold Brook Farm, Stowe, Vermont, March 1947. The maple tree had blown down during a snowstorm. Photograph by John Eagle.

track, and then after you got your load unloaded, you would turn and make a track back for the next day. But if a big storm came in the meantime you would go out first and make sure. If you were doing a ten- or twelve-acre field, you would probably do half the length of the field in one load.

Willie: The manure job was the job in itself. It might take you as long to do the manure work as it did to do the milkin'.

Mason: Yeah, because you had to load it and take it out, either on the wagon or on the sled and spread it by hand.

Willie: And you had to make sure there was enough slippery stuff on the cement so you could start the sleds out.

Mason: Lots of times you use a pipe to roll it.

We didn't have a spreader. Oh, we had a spreader; we used it in the fall but we'd never used that in the winter.

Doug: Why?

Mason: Oh, because the old man wouldn't let you use it!

Willie: Too hard to draw. And you could freeze it up.

Mason: Jesus Christ. He wouldn't let you use that! Manure'd freeze up in it and then you'd break it when you tried to use it! You pitched it in; pitched it out. You dragged a wagon through the barn and you pitched the manure on to the wagon. If was muddy, why we'd put it in a pile, spread it in the winter. Pitch it on by hand. Pitch it into the manure spreader and spread it again. You got to move it twice!

There were jobs, like working in the woods, which horses did better than tractors. Harvesting wood was, itself, part of the world which was passing. With electrification and the routine use of fossil fuels, most farmers replaced their woodstoves with furnaces. This change made rural neighborhoods dependent on distant energy sources, and it ended a practice of putting local resources, such as dead trees, to local use.

It also led to the loss of one form of satisfaction that that many farmers felt. When Mason and Willie discussed working in the woods, they described work that was interesting and enjoyable partly because of the teamwork they shared with horses. It was a sentiment I never heard farmers express when speaking of their tractors.

Mason: We used horses to cut wood (fig. 26). Me and my younger brother would go back in the afternoon, after we got done with dinner. You know, after chores and everything were all done. After it froze up, we would start cuttin' wood with a crosscut and the axe. Probably one o'clock when we'd get back to the woods. We'd cut 'til . . . well, it got dark about 4:30, so maybe four o'clock we'd start back to the house. We had quite a ways to go. Next day, we'd go back with my father, with a team and sleigh. And we'd load the load that we'd cut the day before. And while he was gone with that load, up to unload it, we would cut another. And he never caught up to us. We could cut it faster than what he could draw.

Willie: And you helped him load each time too. Take small pieces of trees—and lay on the bunk and then take your hand-hook and roll on the logs. And then, of course, you top out with the smaller stuff.

Doug: One load would probably be what? . . . A third of a cord?

Willie: A hell of a lot more than that!

Mason: Oh, geez, the old man used to load 'em! I'll tell you.

Willie: You'd get about two . . . two and a half cords to a load.

Mason: Yes. I'll bet he'd have three or four cords to a load. We had a good set of bob-sleighs and they was hooked cross-lengthed on the back, so you could go right in around the trees. You never caught your bunk on the tree. Wherever you went with the front, the back one would come right around too. So it'd pivot, in other words. And, of course, we would fell the trees so that when he drove up to 'em . . . the big part of the tree would be towards the front.

Doug: How many horses were pulling that sleigh?

Mason: Two.

Doug: Two and a half cords?

Mason: Oh, at least. 'Cause you take them sleigh lengths, you know, you get a lot of wood on 'em. And, of course, the old man always had high stake racks. And when we got about half loaded, he'd put a chain around those, then he'd put the others on top so it wouldn't bust 'em. It'd be four or five foot high.

You had to be a teamster to run a load like that, 'cause you couldn't let the team take right off straight. You're gonna bust somethin'. You had to sidestep toward the front, break the runners loose before you took off. Your sleigh'd freeze down while you were waitin'. See, your runners are warm and when you stop, they'd freeze. So of course the team knows this after you do it awhile. When they go to start, they'd swing it one way or the other to loosen it, you know.

Willie: They'd sideslipped . . . and pulled at the same time.

Doug: I can't imagine two horses can move that much weight.

Mason: Oh, boy! They can on runners. Yeah. Goin' to the woods it was kinda' downhill but comin' out was mostly uphill. Of course, you had to stop once in a while and give 'em their wind. We had a big team. Probably weighed a ton apiece.

Willie: They were nice!

The transition from horses to tractors made obsolete a kind of work, and a kind of worker. Willie's comments are a small essay on skills and culture lost with the end of the horse era.

Willie: Set of torches there (fig. 28)—there are tanks right up there. There is a torch lying right there in the back of the truck. Cutting torches; acetylene torches.

Doug: What would he be doing with this?

Willie: Well, a lot of them didn't like to do their cutting on the anvil. Normally you've got a wedge chisel you set into the anvil and then you hit your metal over the chisel. So, when the torches came into style, a lot of them used their torch with acetylene to cut.

Fig. 27 Farmer waiting to deliver milk at Evans receiving plant, Cortland, New York, September 1945. Photograph by Charlotte Brooks.

Doug: Did you have to deliver your milk to Madrid?

Marian: Bill Fisher came and picked the milk up.

Carmen: But I drew the last two or three years myself. When we didn't have the truck Bill Fisher drew it. Bill had a route, he picked up all the farmers' as he went up the road . . . not them all, but those who didn't draw their own. At the time we only had eight or ten cans.

Doug: Eight to ten cans of milk was enough income to run the farm?

Carmen: We got by anyway!

As you know, my dad was a blacksmith. He had acetylene torches in the thirties. At that time, your acetylene came in granules and you put them in a tank. You added water to get your gas. You didn't get the acetylene already made up in a tank, like you do now. That is what this tank is, up in here. It is not a high tank, or a long tank or anything; it is just a round tank you put your acetylene in. The last one of those I seen was Jack Rickett's up in Ogdensburg.

. . . He wouldn't use those torches horseshoeing. You could cut your metal that you use in the knicker shoes with it. Other than that, no.

Looks to me like he has a horseshoe laying there by the firebox, near the forge. Doesn't that look like a horseshoe laying there?

Doug: Well, there is a horse here; possibly he has already shod the horse. Would he go out and fix farm machinery with that rig?

Willie: Oh yes. Same as I do; he travels wherever they call him.

Doug: What would he use the forge for?

Willie: Well, that is one of the ways they had for heating and making parts—hammer parts out—because a lot of your parts on the old wagons and things were hammered out anyway.

Fig. 28 Traveling blacksmith shoeing horse in stockyards, Lancaster Stockyards, Lancaster, Pennsylvania, September 1947. His forge is especially installed in the back of his pickup truck. Photograph by Sol Libsohn.

Doug: Could you explain how you weld on a forge?

Willie: Well, if you were using flat metal, you would hammer the ends of it into a wedge shape. Heat it and hammer it, and you have got to lengthen it as you hammer it out, to overlap it. Then you get both pieces—that you have hammered to a wedge shape—to a certain degree of heat to weld, and you apply a flux to help them meld as you hammer them together.

When you got them both to a certain temperature, you sprinkle the flux on them; hammer them and weld it.

You know when it is the right temperature by the color. Something between a cherry-looking hot and a white hot. Kind of an in-between color, but it depends on the type of metal you are working with, too. Some metal you've got to get hotter than others. You have to have a knowledge of the heat, and you don't learn that by yourself. A lot of the time, you have to learn from someone else, watching.

Doug: What parts would you be fixing, for example, on a plow?

Willie: Well, you've got your colter braces, things of that nature. You'd weld them back. Usually they made new ones—they didn't do much welding. And like your old-time pumps, there with the big handles, you use to weld the shafts and things for those. They had the suckers way down inside the well.

Doug: Did he have to put extra springs on that truck to carry the forge and the anvil?

Willie: That looks like it might be a one-ton or three-quarter anyway. It's well built in the box and everything; that is your bigger style. Your smaller truck doesn't have a box built up that good. A three-quarter-ton, that would haul everything he has got in there.

He'd fire his forge with soft coal; start it with any type of dry kindling wood—most of them liked to use cedar or chestnut.

Used bellows to get the heat up. This is bellows right here on the left side there. Looks to me like he has the crank there in his hand right now. Turns that; turns a fan in this bellows here—blows the air in and you get your heat according to the speed you turn that fan.

He is blowing right now, the way those flames are; he is cranking the bellows. Can you see the fan inside of that casing?

That is the type I had when I had one set up here. The shroud is made out of sheet metal, because all your heat is down in this box. On the average you've got clay around your firebox so it don't get the other stuff too hot. Brick or clay—firebrick. Some of them used firebrick—in them days, firebrick was a little hard to get ahold of, so most of them used clay.

Doug: How thick would that clay be?

Willie: Depends on how deep a firebox you got. [Usually] it would be about three inches on the bottom. The air comes right through the bottom. There was a grate on the bottom like—there's also a place where the air comes through, a shutter, doorlike.

Doug: How long does he crank that bellows to get the temperature up?

Willie: Well if you got good soft coal, it don't take long.

Doug: Thirty seconds—a minute?

Willie: If you've got the fire already going, your heat comes right back in about forty-five seconds or so.

Doug: Does the fire die down then?

Willie: Well when you stop cranking it, it'll die down—smolder like—until you crank it again.

Depending on the work you are doing, some metals you put right into the coals. Like for welding, you have to heat the metal continuously around. If it's a big piece of metal you're welding, you have to turn it so that you have the same temperature all around.

My father had a rig that he traveled with, but it was on four wheels, just like a wagon. It was just like a shed sitting somewhere. Like you picked a shed up and set it onto a wagon.

See, on the average, there was one fellow like this in every community—but there was more than one who could do it. Usually, there wasn't that much work, to put three of them to work. Then a lot of these guys had a young fellow with them as a helper. An apprentice like—that's why so many knew about it.

They usually they started out doing farm work and things, and then went into horses right with it. But a lot of blacksmiths wouldn't shoe a horse . . . you got kicked around!

Doug: In other words, the fact that they were good at the metal didn't necessarily mean they were good with the animals.

Willie: A horse could tell if you were afraid of him; they sense it. And if you were afraid of them they would play with you. If you were rough with a horse they knew it—they wouldn't let you touch them either.

But they always had someone around to do the horseshoeing—a lot of farmers did their own shoe work, but they needed a blacksmith there to make the shoes. You see, horses wear different size shoes, just like human beings do. You have different sized shoes—when you buy a shoe for a horse, you have to know what size shoe to buy. These blacksmiths had dies which they formed their shoes over, and they used to make shoes and keep them on hand. It wasn't just going in and say "shoe my horse"—and make a shoe right there. You usually had some on hand, hanging on the wall to shoe them with—and these traveling blacksmiths like this—if they shoed horses they always had different sizes of shoes to go along with it.

Doug: Did these men become mechanics when it became more typical to have tractors in the thirties and forties?

Willie: They had to be mechanics to be blacksmiths. If something broke on a piece of farm machinery, on the average you were the one who took it off—repaired it and put it back. If you fixed it and it didn't fit just right, you had to keep altering it until it did fit right. So, you were more or less a mechanic as well as a blacksmith.

Doug: What was the status of these men in their communities?

Willie: It was pretty high. They were well known wherever they were. And if you had a real good one, he was known not just in one community, but in quite a few. Some of them traveled.

They made their living and that was about it. Some of them pretty much just scratched a living out. Because the farmers didn't have the money at that time to pay you either.

Doug: See if you can remember what it cost; for example, if he takes his rig out to the fields and fixes a manure spreader for a farmer—what would he charge?

Willie: Depends on what he would be fixing and how much you had to pull apart. He might of got fifty cents or one dollar—dollar and a half.

Doug: He would spend a morning on something like that?

Willie: Easy morning . . . he had mechanic tools with him—wrenches—see, these tools laying right here alongside the torch are blacksmith's tools. Long tongs, and things to pick up the hot metal with, like the ones I've got up at the shop.

The blacksmith wore a vest and a long shirt, even in the summer. You would be surprised the sparks that come off that stuff when you are hammering.

There were some guys—I used to know one—they would take a piece of metal and put it on the anvil and start beating it with a hammer. And they would get that hot just from hammering it. Now, a lot of people think it is impossible—it is not. They get that hot enough and they would start their coal fire with it. We had one guy out here who had a forge—that is how he started his fire. It was easier to start a fire that way. I mean, that metal would get *hot.* You go and start hammering on a piece of metal and see if it don't get hot!

Most of them would use just scrap. Most of them didn't like to buy new stuff.

Doug: Kind of a nice photo.

Willie: I like that photo.

Doug: When I saw that I said that is what you described to me a lot of times.

Willie: I'm trying to see what they had over here. See this box over here? It's a nail box. There are different size nails in here. They might have corks in there to put into the shoes. He has his other tools on the far side of the box which you can't see there . . . the hammer and nips. They called them horseshoe nips at that time.

He's going to shoe that horse; he'll fit the shoe to the foot. They custom fit them; if you don't, they'll tear the inside of their legs up when they are walking; stumble.

Same as you go to a shoemaker to make you a pair of shoes—it has got to fit your foot. It is the same thing. Depends on the horse—the horse's hooves, whether they're close to a standard size, or whether they'd be something you had to make—in between sizes. Some hooves don't run back as far as others. You've got to compensate for that too. And if you put shoes on for winter, you've got quartz in the shoes, which is kind of a spike for the ice. You have got the ice spikes, and then you've got the ones where you pull the spikes out and put other spikes in.

Doug: Would it take an hour to shoe a horse?

Willie: That would be rushing it. Usually you figure on a full morning for shoeing a horse. Unless they have decent hooves. Depends on how much trimming you have got to do. You've got to trim a horses hooves. This guy would also be trimming the hoofs with those nippers. He's got nippers he trims with, and he has a knife he cuts out inside the hoofs with to get the dead stuff out of it.

A horse that has been shod three or four times—they know what is coming. It doesn't hurt them unless they get a nail wrong, and it goes into the nerve part of the hoof. See, the nail is tapered so that they won't come out when you drive them in. I have seen them put a nail in and get the taper in the wrong way, and man, the horse would let them know it—they pulled that nail back out in a hurry. *If* you can hold him long enough! . . .

That anvil's a 125 pounder. About the same as I have up in the shop. He lifts it in and out of the back of his truck, which is what I used to do with mine. But I have never seen anyone use a barrel like that. We used a block of wood. You have got to keep them staves pretty tight on that barrel to hold that and to hammer on it; he's got to have that staved up pretty good.

Doug: Would the father typically teach the son?

Willie: On the average that is what they would do.

Doug: They wouldn't be a farmer *and* a blacksmith?

Willie: I don't know; I've seen some farmers that were blacksmiths. My father was. Blacksmith work was done on the side. There was enough of us that we took care of the fields and things. His blacksmith shop was bigger than my house. He had special stalls to tie horses in to shoe them, and he had the ones that would kick like hell—he had a way of tying them up on posts—half hitch. Run the rope like just above their hoofs. Once you got one hoof off the ground they knew they had to stand there. You usually could do it.

Doug: How old were you when you started helping?

Willie: First time I remember working with him, he had me on a little milk carton crate—standing, turning that forge. I was about five years old. He had to get me high enough to turn it. By the time you were six, seven, eight, you knew something about what it was like, but you didn't get into hammering. Those hammers were heavy for a little fellow to use: anywheres up from one pound to maybe five pounds or more. There were a couple there that were maybe sixteen pounders. Started out with them and finish up with the lighter ones. And then, some were flat on both sides, like that, and some were wedged-chisel hammers. You had chisel hammers and then you had hammers that you cut with too. . . .

That building across the street looks to me like part of an old mill or something. You have an elevator here—it looks like an old grain mill or something. It could be a milk plant to. I bet that's what it is. Milk plant. An old creamer.

Back in those days, they didn't call them milk plants; they called them creamers. You separated your milk—took the cream, just the cream itself—you got a better price out of it.

Arthur: A guy, Jim Coora, downtown, used to do that work. You don't remember that, do ya?

Willie: He's the only one I remember. I'd never heard him talk about it though.

Arthur: Well, there were two or three around. Russ Ellsey used to. Yeah, Russ Ellsey was a good shoder.

Willie: Russ would do it. There was a guy in Madrid use to do it. I remember that.

Arthur: I can remember over where the fire hall is, I think. They would take your horse in and they'd shoe 'em for—oh—fiftly cents a horse. We'd buy the shoes. He'd fit 'em right on the old horse. Cut 'em and trim 'em and [*snaps his fingers*] put the nails right to 'em. This fella with a little portable unit on his truck, well, you see, that was later. That truck would be a '35 or '6, eh? '32?

Willie: No, '35. It's got the straight end.

Arthur: Now they're generally retired. They died. You know, they get old. They'd do the ironwork and make the sleighs. Build the sleighs and put runners on the sleighs. Wood wagon wheels with iron on the outside. Remember they used to make them wheels?

Willie: Remember it? My father was a blacksmith. He used to make the spokes—everything.

Arthur: Yeah. I said to the guy—of course, we were kids . . . maybe sixteen—he said he wanted a iron made to hold a cow's nose. Two holes on each side. I took him down a wagon wheel to use for materials and told him what I wanted and I said, "How much, Jim? I ain't got much money." "Well," he says, "let's see how much it costs." And he fixed it. Put the holes in it. Said, "How'd it work?" Probably worked an hour. Heh? A quarter is what he asked. Well, he was a quarter ahead. Didn't cost him much. In them days if he got two dollars a day he had a lot of money, didn't he?

Willie: Yep.

Willie: He's making a bushing, or a washer, because, see, the hammer he is using is wedged on one end (fig. 29). What he's doing here is trying to cut it down. Bring it out there, then you cut that bunch off. That's white hot. . . . You can see that he's cutting the same size all the way around, and he might be bringing it into that size. You could chisel it right off. I haven't seen any of these chisels or anything lying there.

Fig. 29 Blacksmith, dairy farm, Maple Lane Farm, Greenport, Long Island, October 1945. Photograph by Gene Badger.

Doug: Wouldn't the chisel pick up the heat of the metal?

Willie: You don't keep it on the chisel that long. You put it on there, you cut it, and you get away from it. You hold the piece of metal out over the chisel with your tongs or whatever is long enough to get away from the heat. You hold it right over the chisel, and take the other hammer and hit down over the chisel and cut it.

Your chisel sets into a square hole in the anvil so it can't turn or anything when you use it. If you had a round shaft on the chisel it would go anyway it wanted when you hit it.

He's not using the hammer as a chisel; he is working this out; making a part. Now there's one type of piece like that, that was in the hay-rake wheel. That could be what he is doing; fixing the hay-rake wheel. Part of the hub.

Doug: Lot of patience in that face.

Willie: Oh yes, if you don't have patience, you can't be a blacksmith. Takes a lot of patience, and if you don't get it the first time you try again—it don't always come out right the first time.

Doug: How long does the metal stay hot before you have to reheat it?

Willie: That depends on what you're doing. What you are doing with it. He's putting it over the horn of the anvil there. That's not going to stay hot very long because that anvil horn is going to cool it. He has got to work fast with it.

Doug: He's not going to have a parallel hole in the shaft because that anvil is tapered.

Willie: The anvil is tapered, yes—but, see how he is holding that—he's holding it flat to the top side of the anvil. You can just about see through the bottom side. . . .

Doug: How soft is the metal when you hit it? Do you see it change?

Willie: Oh, yes. The hotter you have the metal, the easier it changes. The molecules of the metal change easier when it is hot. And, when you're finished you temper it. There is water temper, sand temper, oil temper. The idea is, to temper it, you change the temperature radically, but each method of temper changes it differently. You can temper it so hard that when you hit it with a hammer it will fly apart, just like glass.

That part looks like part of the ratchet. The hay rakes had these, and so did your mowers—horse-drawn mowers. That was the part that was made right on the shaft and had a ratchet on it, so that when you backed up it didn't run no more, and when you went ahead it run. If you were making a turn, one wheel would pull it more than the other—run both wheels at the same time. See, your dump rakes had them. After it got up a certain height, she had a little dog on it, and she kicked the dog in and back down. She would go and she wouldn't do that again until you touched the pedal. That dog would kick out and catch it and go around and come to a certain point in the wheel and dump back down again. That is how the hay got raked into rows. But that's about what the ratchet looks like.

Doug: So, when it was worn out you didn't get a new one; you had the blacksmith make you one?

Willie: Well, on your rakes like that, you could get new ones if you could afford them. It was cheaper to have them made.

Doug: It was cheaper to have them made!

Willie: . . . You get pretty dirty as a blacksmith. Sweat all day in the summer; you're around the coal smoke all the time too. Just like a fireman on a train.

Doug: These guys have reputations for being a little different?

Willie: I would say they are more subtle than the average. . . . Little more calm. But don't get them riled up. It takes a lot to get them riled up on the average.

Doug: They wouldn't joke around?

Willie: Oh, they would joke around—yes. Jokes would go just so far, too.

Doug: People probably respected their strength, among other things.

Willie: Just like Earl Crossman—one of the best machinists who ever walked—he liked joking around and things too, but don't get him shook.

Doug: Compared to the farmer though, did he have a different personality?

Willie: Never thought much about it. I would say there was. He always had

Fig. 30 Ed Ruling, eighty-four-year-old harness maker who has made harnesses in the same location, North Collins, New York, for sixty-two years, August 1945. Photograph by Sol Libsohn.

this idea "a little more gentle and you won't break this or that. Don't get out here horsing around with your horses!"

Doug: In a lot of cultures in Africa, Siberia, and China—all over, in fact, the blacksmith is thought to know black magic. Even when you study European history, you hear that people were afraid of the blacksmiths, partly because they didn't understand the science at the basis of what they did, how they changed metal, for example.

Willie: Yes, because they made things and they couldn't understand how it was done.

See how easy it looks—how he's holding that hammer. Not a real grip on it or anything; it's just holding it easy. You don't grip a hammer hard because if you hold a real tight grip on it all the time, you've got a tired hand before too long. You have got to be able to relax when you are hammering and still use it. All day long if necessary!

Willie: That is a great picture. You see a little bit of everything. All you see up here is the strapping for making harnesses. These are tug straps here . . . that is what you hitch up the singletrees with; behind the horse that you hook on to pull the rig. The leather above the guy's head already has buckles on it;

in your spare time, you used to put buckles on a lot of it, so you just cut off what you needed and it would already be buckled. That looks like he has already got straps made up for harness. There is what they call a "britches strap" on the left side of the photo. That lays right on the back; on the top side of the animal's hips, and you've got a strap that comes down around under their tail to hold their harness so that it's not on their back.

That ring clips on the tugs here. They're usually made so that when you're not using the horse for anything—just walking him or something with his harness on—you hung your clips into this. That kept them up out of the way.

On the far left is a bridle—see the bit? He's got it hanging by the bit. You see, he has got a wire across it so it'll hang straight—it don't get all messed up or anything.

These lines, behind his head, are used for securing horses.

Doug: How can you tell?

Willie: They're not very thick, about three-quarter wide. They are wrapped up, and you've got one end up here, and the other end is way down here. He has got an eye to hook on to the bit, and he has got one up here. They will take a certain length of leather, measure it out to make two lines—two lines so when you buy it, you staple and stretch it out and cut it in the middle. So you have got a set of lines right there.

You have got to go according to your horse's size; how far back up on a wagon or a piece of equipment you are going to be sitting. Most of them like their lines long, and then you would tie them together at the end, so you could keep them up. . . . Look how well he's kept—how old did you say he was?

Doug: The caption for this picture was that he had been a harness maker for 62 years at the same location—so if he started when he was 20 years old he would be 82 years old.

Willie: He is a well-kept man. A harness maker can wear a tie!

. . . A blacksmith would usually put a real heavy harness together, because if you had a good heavy team, and was pulling good heavy loads, most of your harness is riveted together instead of sewed. You sewed and riveted, and those rivets had to have washers and everything on them—and you did that right on the anvil.

Doug: So the blacksmith also made up the buckles and stuff like that?

Willie: Some of them did. It was more fine work but on the average you could go into what they called the five-and-ten stores at that time and you could buy a lot of this stuff from them—the old country stores. It was cheaper to buy your buckles and things. They were mass produced.

Doug: The fact that his work was clean and the blacksmith's was dirty didn't affect their respective standings?

Willie: I would say it would run just about the same.

Doug: Did he tan hides?

Willie: If the guy was just making harness alone he would tan hides—that is quite a chore, just tanning hides. But just making harnesses—you were never continuously busy making harnesses, so he had his own little tanning that he used to do. That is why I was looking at the way he has got this back room behind him. It looks to me like it would be a tanning room back in there.

You'd never see farmers wearing a hat like that—a rim hat that would keep dust and stuff out of their necks.

Fig. 31 Shallow discing and rolling field with McCormick-Deering Farmall for planting winter wheat, West Penfield, New York, August 1945. Photograph by Sol Libsohn.

The SONJ photographers recorded the overlap between horse and tractor technologies. Jim Carr, looking at figure 31, said simply: "That was a rough-riding son of a bitch!" The image of the repairman crouching beneath the largest tractor one could buy in 1945 (fig. 32) could be made in any contemporary repair shop of the

Fig. 32 Tractor repair at Kelly's garage, Perry Center, New York, August–September 1945. Photograph by Sol Libsohn.

North Country. In fact, it looks a great deal like those I made of Willie fixing these first-generation tractors, still in use in the 1990s.

But these images raise questions about the appropriate level of agricultural technology. The transition from horses to tractors seems inevitable in hindsight, and most would scoff at a return to, by today's standards, the inefficiency of horses. Interestingly, though, Amish farmers, drawn to the North Country by inexpensive land, have reintroduced horse-powered methods, and their methods seem oddly rational in a setting characterized by the spiraling size of farm machinery and the number of farm failures. The Amish have only a different perspective—the fact that their horse-powered farming is successful and self-sustaining in today's North Country is not a freak of history or social organization.

Most will not return to horse-powered farming; they do not have the necessary stamina, knowledge, or values. Raising the issue, however, calls into question the ever-increasing mechanical horsepower on the farm. When I surveyed forty-eight farms in 1990, the average horsepower per farm was over 400, this being distributed among four or five tractors and often a self-propelled chopper. It is impossible to fix a value on that number, to determine whether it is too large or does not yet realize the potential of a small labor force with a lot of work to do. But the American farmer seems only to imagine a larger machine. In the North Country (as in all northern dairy neighborhoods), short growing seasons and muddy falls compound the normal challenges of getting a crop in and out of the ground. Smaller tractors with four-wheel drive and smaller choppers and other implements, perhaps even used with draft animals for some tasks, are probably, in the overall sense, more efficient.

Or machines could be shared; farmers could again work together in one way or another. European farmers I have interviewed maintain such practices. But Carmen Acres did not believe that the European system would work in the United States:

> *Carmen:* The French farmers our daughter Mardi visited shared their hay bine, and their machinery, their tractors and stuff. Two or three other families. He had a hay bine and Mardi said that they were waiting for the neighbors to bring it back. They did share the machinery—they didn't have the machinery around that we have. I could see where that would work all right if you could get along with it. They exchange with one another over there some. I gathered that there were about two different farmers that worked together with them. Over here, I don't think you could get along that way! That's my estimation.
>
> *Doug:* Why not?
>
> *Carmen:* Well, we need our own equipment! Over there I don't think they do so many acres, and I don't think that it would work—like if I was sharing a machine with James and he had the machine and he brought it back and the sections were all off it—you'd have to fix it and there'd be some confusion there. You'd have to have something set up so that when they brought it back it was in good working order, the same as when they took it. I don't know if you could get along that way or not. I think it'd be hard. Or else someone would drive the hell out of it.
>
> I could see that if they had thirty-five or forty cows and the neighbor had the same it would almost seem as though one farmer, or two farmers, could own a hay bine, and the same, as we used to at home—I don't know—we never had any trouble. Of course, my father always looked after the tractor—it always stayed at our place and the threshing mill always stayed at the other guy's place; he stored it. Dad was always the one who looked after it.

Jim Carr, however, saw the practices of the local Amish as a wake-up call:

> With the Amish comin' in—well, land is cheap here and so maybe it'll bring some change. Now those people are workers. You can't take nothin' away from 'em. You know? And they make a success of somethin' like this; a lot people say, "Why, them sons of bitches, they should stay where they were born." But they see an opportunity and they do it. Through hard work. And you can't explain it to the neighbors, you know?

Making Hay

Dairy farmers, until recent miscarriages in the organization of agricultural work, have had two sets of tasks. One is animal work: breeding, feeding, and milking. The other is growing cow food. Animal work takes place all year long, though at different rates; crop work happens in short bursts of intense activity: planting, cultivating, and harvesting. While farm families generally assume responsibility for animal work, harvesting often requires additional workers. This is a perennial problem of agricultural work and the topic of much of the following discussion.

The simplest way to make milk is to have cows graze on pastures of natural, mixed breeds of grass, which reseed themselves, and some, like clover or alfalfa, which replenish the soil with nitrogen, the primary nutrient extracted during plant growth.[10] Eventually, speaking of long historical eras, farmers began to plow and replant fields to achieve one purpose or another. They might rotate crops to improve the soil, or plant grasses which produced more or matured earlier.

Farmers harvested hay meadows with a scythe, raked the hay into windrows to dry under the sun, and then gathered it, forked it onto wagons, and moved it to a barn to be stored for winter feed. For thousands of years, this was essentially how small farm harvests were organized. It depended on a few tools—scythes for cutting hay and rakes—wagons, animals to pull the wagons, rudimentary barns, and family labor.

While labor intensive, it was ecologically reasonable, because naturally grown legumes replenish soil nutrients needed to grow plants. Mixed varieties of natural grasses are resistant to diseases and insects. Because of minimal plowing, topsoil was not exposed to erosion, compacting, or rutting. But, in its simplest form, it was an inefficient system. Pastures were overgrazed, and farmers lacked the knowledge of soils and plant needed to maximize production even of grasses. Dairy cows struggled through winter

Fig. 33 Hay-baling machinery on a farm in the valley south of Rutland, Vermont, August 1949. Photograph by Charles Rotkin.

Fig. 34 Newton Bush, H. O. Kramer, and Jacob Kramer, dairy farmers, Erie County, near Buffalo, New York, July 1946. Photograph by Gordon Parks.

months because farmers were often unable to grow enough food to last between growing seasons. Most cows were bred to bear calves (to *freshen*) in the spring; thus, farmers stopped milking them (*drying them off*) six weeks to two months before they gave birth. Times were not plush for either cow or farmer in the late months of the winter.

By the mid–nineteenth century in the United States, dairy farmers had begun to grow several crops as feed supplements. These included specialty grasses that matured earlier, which could be fed to freshening cows before the pasture grasses had matured, and grains such as corn and oats, which were harvested in late summer and fall and then stored for winter feed. These became the staple feeds for dairy cows for a hundred years: grasses, both fresh and stored, oats (which also fed horses and oxen), and corn.

The postwar era was a time of great change in harvest technology, the organization of the harvest, and the crops farmers grew.

This topic first drew me to study dairy farmers. Growing and harvesting crops presented the central drama of farming: in my neighborhood I watched farmers on huge tractors or choppers digging up the land in the spring, baling or chopping lush June hay, and chopping huge fields of corn when summer turned to fall. This work was done by individual farmers and appeared to be mostly about machines. But powerful as they evidently were, they were also a terrible headache! The huge tractors and choppers got stuck in the late spring or early fall mud, and even larger tractors were called on to drag the beached behemoths out of the muck. When their enormous power strained their parts to breaking, the damage was severe.

Every farmer owned his or her own equipment, and many of the machines were used for but one operation and then put away until the next year's work called them forth.

When I got to know my neighbors I learned that farming had not always been this way. Farmers had cooperatively harvested oats and corn and had done other jobs together in winter months. Noah Blake, the previously mentioned diarist who lived on a farm in 1805, said his family "changed works" with their neighbors to harvest their hay. Eric Sloan writes that "this was a custom practiced religiously in the early farm days; when there was a job that took more than one person, a group of people did the work, not for pay, but to get that same kind of job done for themselves later. In that way, there was no hiring of help. For example, when a harvest mowing job was ready, some eight men showed up at the designated farm and did the whole job in one day. Then each of the eight men was entitled to the same service on his own farm. Food and switchel (drinks) were the obligation of the farmer whose land was being worked."[11]

The practice of working together for collective tasks is part of most agricultural systems which are not entirely dominated by machines.[12] In the North Country, during the era photographed by the SONJ photographers, farmers changed works to harvest oats and corn and throughout the year for occasional jobs. During the end of the SONJ era, during the mid-1950s, this system disappeared when new agricultural technology came into use.

The shared labor and the activities which surrounded it both created and sustained the collective consciousness of rural neighborhoods. Durkheim noted the importance of a shared physical environment in this process:

> In a small society, since everyone is clearly placed in the same conditions of existence, the collective environment is essentially concrete. . . . The states of conscience representing it . . . have the same character. First, they are related to precise objects, as this animal, this tree, this plant, this natural force, etc. Then, as everybody is related to these things in the same way, they affect all consciences in the same way. The whole tribe, if it is not too widely extended, enjoys or suffers the same advantages or inconveniences from the sun, rain, heat, or cold, from this river or that source, etc. The collective impressions resulting from the fusion of all these individual impressions are then determined in form as well as in object, and, consequently, the common conscience has a defined character.[13]

I feel this sense of collective consciousness in Gordon Parks's photograph of three farmers in a moment of leisure (fig. 34). It is seen not just in their matching work outfits, but in their similarity in posture and expression. Buck said this about the photograph:

> These guys look like rugged individuals, you know what I mean? The farmers today—there are a lot of rugged-looking farmers today, but then there is also a new generation of farmers coming in, for example, like Kevin Acres—now Kevin is *neat.* He could pass for anything!
>
> Do you know Steve Tiel in Lisbon—he is the one in the legislature? Well there is another farmer that could pass for a businessman, or he could pass for really anything. Where years ago your farmer looked more or less the same.

Arthur looked at the photo and simply said:

> They've all got suspenders on. Geez. Now, you see the age . . . the age.

The photos and interviews that follow explore this the collective conscience in the details of shared work. For some of the farmers, these conversations brought disturbing realizations of what had been lost. It is true that traces of what it meant to be a neighbor survive, but the essential shared culture—the collective consciousness—disappeared with the arrival of the machine in the garden.

Making Hay

In the early 1950s, the farmers I interviewed typically had a one-hundred-sixty-acre farm with twenty acres of oats, ten to twenty acres of corn, and thirty to fifty acres of hay, feeding twenty to fifty cows. The system changed as farmers bought new machines and adopted new cropping practices. These changes began a reliance on purchased feed, fertilizer, and other chemicals. This was to be paid for with the increased income from larger herds.

Among the most important changes came in the harvesting of hay. My subjects tell us how the work was organized, and they remember what the work meant: what it demanded and what it provided.

The hay harvest typically began around the beginning of July in the North Country. Most farmers owned horse-drawn mowers and rakes (fig. 36) and, of course, horses. A farm family usually had enough workers to cut, rake, and draw the hay to the barn. Willie remembers:

> Typically, a family would do their own hay. You see, when you've got your hay ready to go in, you couldn't wait for somebody to go in and help you. You went out and did what you could yourself—your own family. If you had a hired man, he helped—that is the way you done it.

Nearly all the farmers I interviewed, however, said that neighbors helped each other harvest hay on an informal basis. Jim Carr:

Well, we always done our own and then, you know, if a farmer got sick or machinery broke or something you go help 'em.

Mason described a similar system:

If we got done first, we'd go help somebody else finish. Or if they got done they'd come and help us. It was all worked back and forth. There was no money involved. Of course, nobody had any money!

Kids helped; farmers hired high school kids on summer vacation. Haying became a family affair; there were jobs for younger kids, perhaps even driving the small tractors of the SONJ era. When I worked with Carmen Acres, one of the first jobs they gave me was raking windrows of hay with a 25-horsepower 1949 Ford tractor, turning it so it would dry evenly. It was a job which the least skilled or experienced could take on. I enjoyed the relatively mindless hours of driving the small rig through the fields, stirring up the hay to dry in the sun.

While much of haying consisted of simple, heavy lifting, the farmers I spoke with recognized the skill it required:

Willie: That was quite an art in itself, just hayin'. They just didn't go out into the field and grab a fork and say, "We're gonna do it." You had to know something about it!

Arthur: You'd get a new guy, and he just wasn't much good to you, was he?

Willie: Not until you got him broke in.

Arthur: You know, I think they had more skill then than you got now . . . hand skill. I don't mean a skill to run a machine, but you had a lot of *hand* skill.

Doug: This became obsolete as soon as you got balers?

Arthur: Yeah. You couldn't get anybody. I wouldn't go help you if you was doin' kile [loose] hay. Oh, I might . . . if it was a little and it was clover hay. Because I used to like to do that clover hay. Didn't you?

You'd pick it up . . . stick your fork in it. . . . I ain't gonna say it'd break off, it'd pull off. You set it on top there and it'd always make a nice kile, wouldn't it?

Willie: Yep.

The SONJ archive offers photos of most of the steps in the haying process. Horse-drawn mowing, as noted above, is pictured in figure 24. The following photographs show important aspects of haying: the horse-drawn rake (fig. 36), loading a wagon in the field (fig. 37), building a stack in the summer (fig. 38), and moving it to the barn in the winter (fig. 42). Libsohn and Rotkin photographed the mechanical hay-loader (fig. 40) which preceded the baling system. Libsohn showed how hay was put in the barn (fig. 41), and Libsohn and Brooks photographed the wire baler (fig. 43) and the machine which soon replaced it, the mobile twine baler. Todd Witlin's image of a

Fig. 35 Shocking hay on a Sussex County dairy farm, Sussex County, New Jersey, June 1947. Photograph by John Vachon.

hay baler at the state fair (fig. 45) marks the state of technology at that moment. These remarkable photographs show three overlapping systems: loose hay, wire, and then twine-baled hay. What we do not see so clearly is the replacement of workers by these ever-more-expensive machines, each of which separated the farmer more from his or her neighbors.

Virgil: I grew some corn when I started here in 1935, but mostly it was hay. We'd fork it into the mow and fork it down. Horse power, of course. That first year we come here we had a barn full of loose hay and I had eleven stacks in the fields. Put it up by hand. Raked it, bunched it up by hand. The bunches were so thick we had to drive over the first row. We'd cut it with a mowing machine drawn by two horses.

You'd take each one of those windrow piles, fold up on each end—that made a bunch of hay. When that got dry, we loaded it into the hayrack, four

Fig. 36 Carpenter Hill, Dryden, New York 1945. Photograph by Charlotte Brooks.

forkfuls on half of the rack. Then you'd put the front on, and then you'd go back to the next back half and put another round on . . . so we had four forkloads. It'd be about eight or nine feet high. We'd draw it off with the horses. Into the barn; back out for another load.

Willie: They used to stook the hay like this (fig. 35) and put it in . . .

Arthur: Kiles, wasn't it? You take a forkful of hay that you'd raked; and the next forkful . . . you pick it up and you set it on. About the fourth forkful you set 'em all on and then you take a fork and rake all them edges out. Right? And if you was good, and it looked like rain, and it was clover, you could have a pile four forks on it maybe. . . . You just take *some* of each kile. . . .

Willie: Yeah, you didn't take it all; if you did you was replacin' fork handles. . . . Oh, you'd break a fork!

Arthur: . . . the rain never hurt it a bit if you kiled it up good.

Willie: If you didn't set it up right it was hard to pitch onto the wagon.

Fig. 37 Bringing the hay in from the field with a small "Cletrac" tractor, Daitch farm, Roxbury, New York, 1945. Photograph by Sol Libsohn.

Arthur: Oh yeah. It'd fall off on you. My father was pretty particular.

If you stooked it right, the water'd run off of it. It wasn't cut or anything . . . only at the bottom. You have to have an awful rain that you couldn't draw hay the next day the sun come out!

After you windrowed it, you took the horse and one of them dump rakes like in that photo. Then you rake 'em. My father said "rake 'em straight . . . dump 'em straight."

Willie: Depended on the horse you had, too.

Arthur: We had a fast horse, and then we had a slow one. He sent me down the road towards the neighbors to rake—ordinarily he didn't swear much. This horse would go just fast enough that the goddamned rake would drop right down in the windrow. . . . And then you'd swear; you wouldn't have room enough to take it all to the next windrow!

Willie: There is an art to building a load of hay . . .

Fig. 38 *Left to right,* Bullard Shafer, Arthur Shafer, and Philip Berg rest from haying in the middle of the day, Fulton, New York, 1945. Photograph by Sol Libsohn.

Arthur: They're pitchin' that hay (fig. 37). See, he's got the back end loaded. That's the way we used to do it. You'd put it in bunches. When you're pitchin', you've got somebody up on top here who's loadin' it and unloadin' it . . . and the other fella pitchin' it on. Remember that Willie?

Willie: Do I? Heh . . . heh . . . heh . . .

Arthur: Give me the next one.

Doug: Not so fast.

Arthur: What else you want to know?

Willie: Explain to him how you had to put that on the wagon so it would stay.

Arthur: Well, you see the sides on this wagon? They call them a coon rack. It would be three foot high. And if you pitched it on, you put one forkful over there, and one over there, and one over here, and two in the middle. And them middle ones are very important because that binds 'em up so they . . .

Willie: Tied 'em together.

Arthur: Just any dummy couldn't load a load of hay. First thing you'd know you'd lose it and he'd have to pitch it up again. . . .

Willie: First little grade of the land you went into—the whole thing'd go right over.

Arthur: Unloading it you took the back forkful first. And then you go from the middle. You see, lots of times you'd have some left. We'd say "hangin' around"; hay that didn't go on so you'd kinda' rebuild that fork. Heh, Willie?

Willie: Yep. And if it'd start to gettin' too much out over the sides, the guy on the ground would strip it down and throw it back where it belonged.

Arthur: Hey, it was a lot a lot of work wasn't it?

The Hired Man

I detour momentarily to discuss the role of the hired man, inspired by the hired man in Gene Badger's photo, likely a key part of the hay harvest.

Hired men were only common on the larger farms of the North Country. On these farms, because of the stability of the local agriculture, and the relative prosperity of North Country farms, they had more privilege than most hired hands experienced during most periods of American agriculture. In most farms I studied, the hired hands were long-term employees who eventually earned enough to buy their own farms. Jim Fisher's farmstead included a handsome stone house one lot from the homestead, which was used for the family of the hired help and is now owned by Jim's son and their family. On the other side of the Fisher homestead is a farm run by the son of a man who accomplished the rare feat of saving enough to buy his own farm. Willie's reflection on this photo suggested that hired hands were integrated into family life, though many remained on the fringe of the system.

Doug: Did the hired men live with the families that they worked for?

Willie: Oh yes, quite often.

Doug: Would they be from the neighborhood?

Willie: Sometimes you would have people come in even from out of state. People that have been around for years, that go from one place to another. More or less just to make a living. They could never get enough to buy a place of their own.

Doug: It seems like a hard job for not a lot of reward.

Willie: Well, in those days there wasn't too many jobs. You took about whatever you could get. He worked for twenty-five–fifty cents a day. Moved right in the house with the family. Gave him a room and done his laundry for him—that was about it. When it came dinnertime they went in and sat down right with the others. If someone was around long enough they became something like an uncle. Some hired men'd go to work at a place and they worked there all their life.

Fig. 39 On the Vedder farm, the hired hand takes time off at noon to have lunch on the running board of his 1929 Ford, Watermill, Long Island, September 1945. Photograph by Gene Badger.

Doug: How is it today? It is different today?

Willie: Altogether different. You go to work in the morning at the farm and you go home at night. You don't stay at the farm. Some farmers have their own tenant house, or a trailer, but not many.

But in those days the guy wouldn't stick around long if it wasn't to his liking. A lot of your hired men at that time were people who traveled the rails. If they could stop somewhere and get a decent job, they would stay—if they couldn't get a decent job, they kept on moving. At one time, men was hard to get a hold of; a lot of times because they just didn't take anything anymore. It is altogether different now.

Each one had his own personality. Just like hoboes—they were all proud. They weren't going to cater to no one.

Taken more broadly, the question of hired labor is part of the discussion of mechanization documented in the SONJ photographs. As we have noted earlier, both the Caterpillar tractors capable of pulling the room-sized combines used in the western grain fields and the 25-horsepower tractors that replaced the two- or three-horse teams of the North Country had been developed many years before they were broadly adopted. The historian Corlann Gee Bush estimated that the typical tractors, such as were used in the North Country, each replaced three to five workers, and the large tractors used in the Western grain harvests each replaced as many as nine workers.[14]

It is also important to see the hired hand as a part of historical change. Tractors were available in the twenties, but by the time the depression reached its full fury ten years later, few farms had amassed the capital to take advantage of them. The depression, of course, released several million workers from the ranks of the employed. Many of these hit the road in search of work, and tens of thousands rode the rails to seek work in the wheat harvests in the American Midwest. Nels Anderson, an ex-hobo sociologist writing during the Great Depression, suggests that it was the threat of strikes by radicalized itinerant workers that drove many farmers to buy tractors and other machines that would lessen their dependence on the migrating workers.[15]

Eliminating hired workers also lessened the farm women's labor, for it was the wives who cooked their food, provided their board, and also washed, repaired, and sometimes even made their clothes. In Willie's comments above, the woman's work involved with the integration of the worker into farm family life is assumed. The history of farm women, however, tells a different story. Taking care of farm hands and integrating them into family life was an often-mentioned irritant. Katherine Jellison reviews a 1940 survey of rural communities and notes that "In the two decades preceding introduction of the tractor in Shelby County [in northern Iowa], women's care of hired help accounted for one-quarter of total expenditures for farm labor in the county. . . . By decreasing the need for hired hands, adoption of the tractor lessened the farm woman's daily labor."[16] Edith Bradley Rendleman, who was born around the turn of the century and lived on farms in Southern Illinois all her life, writes, for example: "That is one thing I hated, to always have somebody around to cook, wash, and iron for. You never got to enjoy your own family alone."[17]

Finally, it was precisely this reserve army of workers that was sucked into the mobilization for World War II. Mason recalled that farm wages increased from less than a dollar a day before World War II to a dollar an hour during the war. Canine reports that the wage pressure led his grandfather to adopt machines to replace his son, drafted into the war.[18] In Arthur's view:

> . . . the war was the end of the whole of cooperative farming. Labor got scarce after the '40s. Labor got scarce because you could go get a job. The boy left the farm and the old man died. The boys didn't want to work as hard as the old man did.

In the North Country, these forces of history and technological change were transforming the social existence of the man photographed by Badger.

The Increasing Mechanization of Hay Making

Fig. 40 Gathering hay with a John Deere hay-loader and tractor on the Gholki farm, Black Horse Road, Warsaw, New York, August–September 1945. Photograph by Sol Libsohn.

The machine pictured in figure 40 was an intermediate step in the evolution of hay-making machines. The farmer had cut the hay and raked it into windrows, either by hand or with a mechanical rake. The hay-loader was pulled by horses or a small tractor, licking the cut hay off the ground and dropping it onto the wagon. Workers caught the hay as it dropped off the loader and stacked it on the wagon. For the first time, making hay was done to the pace of a machine.

The hay-loader, in turn, was rendered obsolete by the eventual invention of the mobile baler.

Arthur: That's a rope loader (fig. 40). Oh my. You cut the hay and rake it in the windrow, and then drive the horse or tractor or whatever you used to pull it. Probably horses, but with a wood-wheeled wagon. You got to straddle the windrow. And pick it off like that and throw it in a pile—two piles. A front and a back so you could take it off. You can't take it all up the whole wagon, so you make the back and then you put the front on.

Buck: You could stand in the middle of your load—a lot of people had a man in the front and one in the back—I always done it alone—and take a forkful and place it right around the edge of the wagon. Of course you had to tramp it too. You know but I could build a load of hay that was just as square as a dollar. You went about six fork high, or about three or foot above the standard.

Mason: Oh, that's a man killer!

Doug: How so?

Willie: Well, it would depend on the guy that was runnin' the wagon. If he took his time and was decent, or if he didn't! Had a guy used to pull one with a ton and a half truck—haul a wagon behind it. We'd load the truck, then we'd load the wagon. You could never get him to slow down. Finally we started breakin' the load. "Keep that up," he said, "and I'm gonna have to fire ya." I said, "You drive that truck right and we won't have this problem!"

Mason: They're putting loose hay into the mow (fig. 41). You had a horse on the other side of the barn to pull that up. And then when it hit the track, it went on a rail, like out of a railroad, which was up in the peak of the barn. There's a guy in the in the mow, at least one.

Willie: That was one of the hardest jobs!

Mason: That was a hot job, mister, under them tin roofs. I'll bet it was a hundred fifty degrees in there! You moved it around to keep it level. . . . You get more in, that way, leveling it.

Willie: Depending on how you mowed it too affected how hard it was to get it out when you was feedin' it.

Mason: Well, if you mowed it away pretty good it wasn't too bad comin' out. But if you just rolled 'em in, them big chunks were a little hard to get out.

Fig. 41 Putting hay up in the barn, O. J. Griffin farm, Roxbury, New York, 1945. Photograph by Sol Libsohn.

Willie: Yeah, you had to take a hay saw . . . fork or a knife, and cut it.

Mason: I done that too. It brings back memories.

Doug: I've got photos of something you've never done any of, which is haying.

Jim Carr: No. Ain't done any since last summer! It's comin' fast again, huh? . . . That's the way we used to do 'er . . . see, your wagon wasn't as long as you have 'em today, so you built your load so that two forkfuls would make a layer.

You'd generally draw it off with horse on the other side of the barn. You'd have a pulley up in the barn—you'd pull it up! There'd be a piece called a "diamond" at the end of your track. Your fork goes up in there, and it locks. And when it locks, then your hay fork'll run along your rail or wooden track in the top of your barn. Ours has a wooden track in it. That's why we had to quit, because the track kept breakin.' And it's pretty hard to repair it up in there, so that's when we had to get the hay elevator.

See this rope he's hangin' on to? When you got it in the barn as far as you wanted to, you pulled that and straightened your tines out, and it'd drop it. And then you had a man up there who'd level it off.

Your load depends on your hay too. If it's straight Timothy, it won't hold on the fork as good as clover or something like that. And with your loose hay like that, if your hay was a little damp you could get away with it in the barn. Because when you mowed it away, if you had a chunk that was a little damp you'd just throw it off to the outside of your mow and it'd dry out. Where if it's in a bale—if it's damp, it'll mold it and there's no way you can prevent it. That's why it's got to be a lot drier to bale. If it's wet, the bales'd be so heavy you can't lift them. Yeah, so you'd know. A lot of people, if it was a little damp, they'd sprinkle salt on it, up in your mow. I've seen my father do it lots of time. If he figured it wasn't gonna keep, just go up there with a 5-gallon pail of salt. And after he'd mow the forkful away he'd just take the salt and spread it on it, and that'd draw the water out—dry 'er right out.

You didn't put as much hay in the barn because it's got a lot more bulk when it's loose. We'd put in about seventy loads of loose hay and then we'd have probably twelve to fifteen acres of hay left over. We'd hire a guy with a baler to bale it. We'd draw the bales and put 'em up on top of the loose hay, and then feed the baled hay out first and then the loose hay. But, my God, after you'd put that baled hay up on there it'd press that down. Man, that stuff pulled out hard!

Buck: We used to put what we called six "horseforks" on the wagon. Six horseforks full made a nice load, and it used to take sixty—three of them loads of loose hay to fill this barn full. Well then, when the balers came out, people had trouble with their barn ceilings falling down because we had it figured that six horseforks was equivalent to fifty bales, so when we put in thirty-three hundred bales of hay that was equivalent to sixty-three loads of loose hay. But the barn wasn't half full, so people would fill the mows right up and the ceilings couldn't hold it.

Fig. 42 Farmer hauls hay load on sled across field, Fish Creek Road, February 1947. Photograph by Sol Libsohn.

Jim Fisher: When the barn was full, the balance of the hay was put in stacks. They decided where there would be a convenient place to make a group of haystacks that would be easy to get to in the winter, so that they wouldn't be snowed under, and they would have access to them, usually near the barn. Probably about eight or ten loads were put in each stack. They were pitched off the wagon by hand, by one person, and then there was one or two people built the stack.

You made a stack probably twenty feet across at the base. Then you kept going up, and when you got so it was too difficult to pitch off from the wagon to the top of the stack, you topped it off to a cone shape. If it was properly built, it shed water very well, so that when you came to move it in, in the winter, it wouldn't be spoiled down over a few inches. Then the process was repeated; you would drive up to the stack—one man on the stack, one on the wagon—pitch it off, build a nice load on the wagon, and bring it to the barn—unload it with the horse fork!

If you had a large hip-roofed barn you might not have to stack any hay. But if it was the old type of barn you would probably be moving hay from February on.

Loose hay gave way to hay baled with machines that compressed it into cubes and bound the cubes with wire or twine. Baled hay was denser than loose hay and could be handled more easily. Baled hay could be sold and shipped to faraway markets; thus, baled hay made it possible to raise cows far from where their food was grown. With this change, the rationale for the northern small farm itself became vulnerable. The connection between farming and the utilization of local resources was broken.

But Willie and Arthur mostly remember what the balers meant to the daily work on the farm:

Arthur: There were balers before most farmers had tractors; I'd say '37 . . . '8. Tractors were steam. The balers ran off the power takeoff from the tractor; well, maybe a motor on some of 'em.

You hired somebody to come in and bale your hay a lot of times because you couldn't afford to own the baler for what it cost you to bale it. And then you had to have some know-how.

Willie: Now, the first balers were "jump balers." They set it in the barn; then they pitched the hay in and they got it full. Two or three guys would. . . . Two guys, right?

Arthur: If you *had* two guys. One guy, it worked the hell out of him. Oh, it'd kill him. Usually there would be two guys together—they jumped the hay down in and that packed it. Then they'd put in some more hay and they'd jump again. When they got it full, then they'd hitch 'er to the horses . . . start the horses up and run 'em around and round. They called that what? "Round the winch?" They would run around and 'round, and that thing would come up make the bales. And then they'd wire 'em. They'd more or less pack it together; it just squeezed it right up. The horses were connected to a winch by a cable—

Willie: —Some of 'em had a rope on 'em, and some cable.

Fig. 43 Emerling brothers baling hay on Lee Nichols farm, North Collins, New York, with Ann Arbor baler drawn by a Massey-Harris Super tractor, August 1945. Robert Emerling feeds and hooks baling wire on baler. Photograph by Sol Libsohn.

Arthur: Then they'd open the chute and out she'd come.

Doug: Was that for selling and transporting, or for storing hay for your own farm?

Arthur: No, no. That's for transporting, most of the time. *Because it cost money*. Yeah, you'd never use that hay yourself! Always used loose hay. . . .

Doug: Okay, so jump bales are first, and then the second stage was what?

Arthur: Portable machine baled 'em; mostly custom? Quite a while before World War II. Yeah, I'd say '34, '37, '38. We hired our hay baled. We had a power press too. That was about, oh, . . . '39 . . . '38. They was very similar to this one here (fig. 43), only you pitched the hay in. Instead of it pickin' it up in the field you'd feed it in, and the old plunger'd go vroom, vroom, vroom, vroom, vroom. Then you'd put a board in and you put a nice forkful of hay into the board, where the plunger would go. And then you'd plunge that until the bell rang. . . .

Willie: Pressure bell.

Arthur: The wire would ring the bales. And then when the bell rang, why, you would put in a nice forkful of hay . . . tip the board. You put the wire back through and then hook it. Am I doin' all right, Willie?

Willie: And the wires had an eye and a hook in 'em. So one end was a hook and one end was an eye.

Arthur: If them fellas was good you worked your hands right off. Then when they come out the back end, you piled 'em someplace. Five bales high. A hundred twenty-five to a hundred thirty-five pounds each. Three men on the baler. And one to feed it. You had three. I got $1.75 a day. Seven in the morning 'till five at night. . . .

Willie: That was big money.

Arthur: Boy, that was good money. He told us if we got twenty ton out, that was a day's work. Not many days you would get it . . . because the guys in the mow never could pitch it good enough. Well, I got up to that once. Old Austin fed the baler. Three o'clock we had twenty ton of hay loaded. And the Frenchman shouted in, "What are you doin'?" "Well," I said, "Twenty tons. A day's work. Shut her right down now." Oh, they'd do that to you wouldn't they? Your word wanted to be good!

Doug: Who'd be the guy who'd buy that baler?

Arthur: Oh, some farmer maybe that had a little money. Depended who had money.

Willie: Or someone who was in good standing with the bank.

Arthur: They cost . . . I suppose they cost $3,500 didn't they? Maybe?

Willie: Yeah, it'd be right around that.

Arthur: And they'd charge a dollar . . . so much a ton to bale it. Heh?

Willie: Yeah, so much a ton.

Mason: Say, they made some big bales and heavy, mister . . . it'd take two guys—all they could do is lift 'er up on the wagon.

Willie: A guy on each end of the bale and hope it didn't break.

Mason: They went around with a baler. You'd pay so much a bale and you'd go out and get 'em to bale for you. They used to get ten cents a bale and then, of course, it got worse.

Willie: You probably done well to have one. If you had one in the community it was a good-sized community . . . you were doin' well.

Fig. 44 Farmer brings baled hay to Warsaw railroad station for shipment, Route 20A, Warsaw, New York, August 1945. Photograph by Sol Libsohn.

. . . He's on an old car seat (fig. 43). The wire was loose; and the wire had a loop. Remember that? You hooked 'em as they come out of the baler. It tightened that wire up. They called it "lacin'." . . . You didn't want to get your hands caught in it! It wasn't the nicest job. . . .

Mason: Keep your hands out of there!

Arthur: You would be selling hay in the wintertime, when somebody'd run out. In the spring, if you had any extra hay you could always sell it. But, of

Fig. 45 Farmers talking behind hay baler, the Great Allentown Fair, Allentown, Pennsylvania, September 1951. Photograph by Todd Witlin.

course, it sold better if it was baled. But a lot of people bought loose hay too. It all depended on what kind of a stable you had. If you had a wide manger you could use baled, but if you had to put it through a trap door the loose hay worked a lot better.

And, of course, they didn't have these wire balers very long. They came out with the string baler. They didn't have that guy ridin' on the side; it became a one-man operation.

Arthur: Oh, he's sellin' some hay . . . baled hay (fig. 44). Put it in the car and shipping it someplace. I've seen us bale right out of the barn, and a tractor-trailer would back right up and he'd go to sleep in there. Probably driven from Albany or someplace and got up there by seven o'clock in the morning.

We'd load his load and then away he'd go. Other fellow used to bring a load of coal up. Sometimes fertilizer. He'd back up and then he'd sleep all day. Those fellows would never get home until the weekend. You only baled it to transport it. Only.

Willie: Ship it to different parts of the country where—if they had kind of a drought or something like that—they used to ship it. Back years ago a lot of your suburban areas of your cities kept livestock, and they used to ship this into the cities, and that is where they got their hay to feed their livestock and things. People used to keep pigs and some of them kept young steer for their own use. Back in them days it was hard to get meat so they used to ship the hay in. Back down home my father used to haul hay—the loose hay on a wagon, clear into town to these people. Never had any trouble getting rid of it. They would run out and fight over it.

Doug: At what point did people start baling their own hay?

Arthur: Forty-five, maybe 1950?

Willie: Right around '45 they started and then more and more got into it.

Arthur: Oh, it was in '48 there wasn't many balers around here. Stanley Salton and I wore out a brand new Holland baler in one year. It wasn't worth nothin' when we finished.

Willie: Continuous travel.

Arthur: We moved all the time.

See, my barn burnt one year. I went over to Stanley Salton and bought a baler like the one there in the picture (fig. 45). I was the only one around here. It was nearly the last of August before we got to hayin' anybody. We had put in some loops and then they wanted it baled so they could get the hay out quick. It got all dry. Everything was dry. I used to bale nine o'clock in the morning till pretty near five, when Stanley Salton come home, then he'd bale till dark. I've seen us get out forty-eight hundred bales. I'd bale twenty-five hundred and he'd bale the other thirteen . . . twenty-one hundred at night. He'd send me to the farmers out here; he'd go over on the island where the hay was a little better and all in one field, and all we had to do was [*Arthur snaps his fingers*] drive, put in new twine once in awhile.

Doug: What tractor did you use?

Arthur: Just a Ford tractor. Nothin' . . . twenty-five horse, probably, heh?

Willie: Twenty-nine.

Arthur: Twenty-nine. Do you believe that?

Twine baler, they had a motor on 'em. There's a few of them settin' around. And some of 'em had even a hydraulic turn. You just turn the thing a little and you could put more pressure on the bales or turn 'er back the other way for less pressure. . . . Stanley said you didn't make any money balin' too heavy so you'd always get off and look and always keep lettin' 'er back. They weigh up near a hundred, some of them, wouldn't they?

Willie: Yeah.

Arthur: Eighty-five—they ought to weigh seventy pounds, about.

Summary

Harvesting hay was a routine job on the dairy farm. It was the first harvest of the season and took place in the hot months of the summer. Haying involved simple teamwork and a lot of muscle. Families harvested most of their own hay, although neighbors helped when a farmer was behind schedule. On the surface, it was simple work: driving a mower towed behind horses or run off of the mechanical power takeoff of the first-generation tractors. After the hay was cut, it was raked with implements pulled behind horses or small tractors. After drying in the sun, it was picked out of the field, stacked on wagons, and moved to the barn. A competent worker built a stack which stayed on the wagon and was easy to load into the mow. The same skills were used to build haystacks left in the fields after the barns were full.

Cutting and storing loose hay, in retrospect, seems to be a simple anachronism of preindustrial farming. The work involved in moving and storing loose hay was arduous. Yet loose hay had some advantages over the baled and chopped hay, the two forms of storage which followed it. Loose hay could be harvested with simple tools. Workers set the pace of the work (at least until the development of the mechanical hay-loader). While it was physically demanding, it was not hard in the way it is to work to the pace of a machine. Recall Mason's comment that the mechanical hay baler was a "man killer."

Loose hay lost few nutrients in the process of being cut and moved to the barn. When a farmer stored loose hay in the barn, Jim Carr tells us, he could assess the dryness of the hay. If it was wet he could salt it, or move it to part of the mow where it would dry. Baled hay can rot if it is baled wet, and in fermenting it can overheat, smolder, and even catch on fire. Many barns burned when baled hay became common.

Baling hay knocks leaves off alfalfa and otherwise drains nutrients from the plants. Several farmers spoke of their preference for loose hay, especially as Amish immigrants to the neighborhood revive the now ancient custom. Loose hay requires some knowledge and skill, but the farmer is more in control of the process and the product. When hay was baled (and, subsequently, chopped), this knowledge, skill, and control was lost. Haying became mostly machine management.

The first balers, which preceded the SONJ era, used brute human effort and horsepower. Workers jumped into the mechanisms, which began to compress the hay. Horses connected by pulleys and winches squeezed the hay further, eventually into bales. These balers, in use in the 1930s in the North Country, made bales which weighed 175 pounds, about twice what a single worker could handle. The farmers I interviewed spoke about having farmers with balers come to their farm after they had filled their barn, to bale the hay that remained. Some of it might be sold, because it could be moved more easily than loose hay. Other baled hay might be stuffed into the barn on top of loose hay which had been compressed. This often caused problems, because it overloaded barns (the barn on our farm had cracked floor joists from overloading) and, in compressing the loose hay, made it more difficult to unload through the long winter months.

Balers made hay a commodity for a nonlocal market; thus, baled hay led to making milk far from where cow feed was grown. If the cows were fed with feed grown thousands of miles distant, their manure no longer nourished the ground which grew their food. The energy costs of making milk escalated as transportation of feed and fertilizer entered the picture.

Hay-making technologies evolved during the SONJ era. Twine balers were just coming into use, and this was the first step in making individual farms technologically independent. Twine balers made bales which weighed about seventy-five pounds; a farm worker could pick them up, load them on wagons, throw them onto elevators, and take them off the elevators to be stacked in the mow. It is twine-baled hay we see being loaded into boxcars in figure 44. Twine balers were not financially out of reach for successful farmers of the era; during the 1950s, progressive farmers like Carmen Acres were able to buy them and put them to use in their own fields and in the fields of their neighbors. In fact, many small North Country farmers use twine balers to this day.

The SONJ photographs and interviews drip with perspiration. I worked in hay mows as a teenager and have never worked harder in hotter or dustier circumstances. Yet this is but one side of this issue. It was hard work, but it was a ritual of the yearly farm cycle. Verlyn Klinkenberg,[19] a city man raised on a farm, described his return to his uncle's farm to help in haying. He tells us that being bone tired after a day's labor, especially when it is shared with others, is not the worst fate known to humans. The system which replaced it is mostly solitary machine work: monotonous, lonely, hot, and noisy. It removes the farmer from the crop; hay becomes a commodity which the farmer never touches. It is also an incredibly efficient system, unimaginable to a farmer who spent all the Julys of his life in the hay mow.

The Modern System

There are two ways to make hay on the modern farm.

Figures 46–51 show the system which replaced the SONJ-era technology on many farms. This is chopping hay and allowing it to ferment in the silo, making "haylage." This is an extremely efficient method of getting hay from the field to the storage facility, but at a caloric cost which includes the energy required to run sev-

Figs. 46–51 The modern haying system, the Acres farm, 1980. Photographs by the author.

eral large machines. Cows eat a wide range of feed, but farmers disagree about whether a steady diet of fermented corn *and* hay is desirable.

The North Country farmers who make haylage use the self-propelled hay bines and choppers pictured here, which cut, chop, scoop, and move hay from the field to the silo.

In the photographs, made during the early 1980s, Carmen Acres instructs his son Kevin on the plans for the day. Kevin cuts hay with the hay bine, and the mower and chopper pass each other on the field. The chopped hay will be blown into the silo to make haylage, which is not photographed in this sequence.

Recent development of round balers offers another alternative for haying. Round balers make several-hundred-pound bales which are several feet across. They are moved with front-end loaders mounted on medium-sized tractors. The machines are relatively inexpensive, and they harvest hay efficiently. They beg for collective use; several moderately sized farms could share a single round baler, since most of their crops could be harvested in a few short days. Indeed, in France, where I interviewed dairy farmers, it was common to share round balers. In the modern United States, the collective consciousness does not lead logically to that alternative.

Oats and Corn: Changing Works

The story of oats and corn can be told simply: farmers informally cooperated to harvest oats and corn with machines such as threshers and stationary choppers, and then, for a myriad of reasons, this cooperative system disappeared. When farmers changed works, they not only took care of their harvests, but also played out their roles as neighbors. What it meant to be a farmer and a neighbor changed when this system disappeared.

Because of the cost of machinery and the availability of labor, it was logical to share labor and some of the most expensive machines upon which the harvests depended. Farming was not debt free, but in this system, escalating machine costs were not an ever-widening black hole, sucking all present and future earnings into the far beyond.

The changing works era was an unusual balance between mutuality and independence. Because the farmers depended on each other, they were all nearly self-sufficient.

The end of changing works was part of a transition to increasing corn production. Farmers began growing more corn because an acre of corn makes more feed than an acre of oats or hay. This was partly due to genetic improvement, through hybridization, which had increased corn production several times over by the SONJ era,[20] and it was very much the result of machines which were developed during the SONJ era.

The end of changing works was also partly due to disappearing farm labor. As mentioned earlier, World War II drew more than five million men from agriculture. Farm labor had been freely given and expected in return, for, as Mason said, "hell, nobody had any money anyway!" As farmers mechanized, they began to think

Fig. 52 Close-up of grain from the Loeffler farm, Monmouth, Maine, August 1944. Photograph by Gordon Parks.

more like capitalists than neighbors. With expensive machines to pay for, there was less time to help neighbors. A farmer could buy a machine which did the work neighbors had done before, and he could do these jobs for a fee, perhaps enough to pay for the machine before it wore out. The farmer began to see his or her own work more as a commodity from which he'd *better* get a good return.

The shift from the changing works era took place in the postwar social and cultural context. Machines symbolized and actualized an ever-widening version of the American dream. Farmers were perhaps slower than others to accept this new ideology, but, in the end, they did.[21] It may have been the farmer's son, returned from an agricultural college or university, who was the critical element in the transition. The father's traditional methods were obstacles to this imagined "good life." Thus, during this era, natural methods were replaced by artificial fertilizers, herbicides, and pesticides. Farming methods were redefined by bigger tractors, mobile choppers, self-propelled combines, and several-times-larger silos for the new king of crops: corn. The Department of Agriculture defined these changes as an inevitable future for American agriculture; extension agents and rural sociologists were cheerleaders in the same parade to the future.

Of course, farmers embraced the technologies that ended changing works because, at least in the short term, it made the physical work easier. But it also changed work qualitatively. The era which followed the changing works system alienated the farmer from his crops as well as his neighbors. I use the term "alienation" in its original meaning to indicate "separation." Verlyn Klinkenborg writes: "A combine ingests corn, soybeans, and oats, and when full it empties its belly into trucks that dump loads through grids in the co-op concrete. Every now and then a farmer or an elevator hand may dab at the grain crop with a feed shovel. Otherwise, no one touches it."[22]

Klinkenborg's observation is a metaphor for many of the changes brought about by the modern era: the farmer no longer tastes his dirt to judge its health because, more and more, it is simply a medium to hold fertilizer, seed, herbicide, and pesticides. The farmer waves from his tractor to another farmer who may be driving by his field, but otherwise he sees him rarely and seldom if ever shares a job with him. The farmer does not feel the heft of the corn, the dust of the hay, nor does he stack bales of straw, for now machines process the crops from planting to harvest. The farmer is no longer in the natural world, in the sense of feeling the heat of the sun, the rhythm of his own body, or the natural cycles of nature. He is in the cab of his tractor, probably air conditioned, listening to the radio rather than the sounds of the fields.

In the discussion which follows, there is nostalgia for a world lost. This was a world of interdependent neighbors, ritualized social events, and connection to land and crops. I believe they represent a possibility as well as a memory.

The passages that follow explain the changing works system, discuss what it meant to those who were involved, and they tell how it came apart. I take the events in their natural order, that is, through the harvest of oats and corn and the other jobs which followed. I also note the playing-out of gender roles in the performance of these tasks. It is remarkable that the SONJ photographers, working separately, recorded nearly all aspects of these mundane events.

The Organization of Changing Works

Adjacent farms usually exchanged labor. Since farmers, by and large, did not choose their neighbors, they also did not choose their work crews. Likewise, being a farmer in a neighborhood which changed works made the resource of a labor crew available. Jim Fisher described the system:

> When it came time to harvest the grain, everyone helped each other. There would be about four or five wagons and teams. Say we started here. The grain would have been cut and stooked and be standing in the field drying and then when the time came for it to be threshed, Hargraves had a threshing machine and the crew cycled around the whole neighborhood. There would be four people pitching the grain onto the wagon by hand and loading the wagons to bring to the barn. Another man would unload into the threshing machine. There would be probably six farms working together, and then another six in another, organized around another threshing machine.

Farmers remember that they contributed equally to the work even if they had a smaller crop to harvest:

> *Jim* Carr: Now, Frank Short lived down where Stuart lives, and he'd only have five or six acres. But, he'd come and help us.
>
> *Doug:* Did it ever not work out? Did some people not work as hard?
>
> *Jim:* No, everybody got along. There wasn't any of this arguing. And it didn't matter who was first or last.
>
> *Doug:* How did you organize who was going to be first?
>
> *Jim:* It depended on when you got your grain in, you know.
>
> *Doug:* Nobody ever felt like they were getting taken advantage of because they were last?
>
> *Jim:* No, never. Nobody ever actually felt that way.

> *Virgil:* They just kept track of when he was coming and everybody who got together knew they had to thresh at that time. There wasn't any complaint about who got done first. The thresher just came right up the road.

Jim Fisher: The farms were about the same size; Lester Joyce was a smaller farm. And Bill Day's was a little smaller. But not enough so that it made any difference. Nobody quibbled about an extra half day on somebody's farm.

Buck: The crews did one farm per day, eight to ten acres. You took yourself and your hired man, if you had one. One neighbor who had a really small farm—well, you could do his crop in two-thirds of a morning. But he worked along with everybody else—he just loved to be with his neighbors!

Some farmers, however, described how the system sometimes did not live up to such high standards of selflessness and equality:

Doug: Were the ten farms who cooperated in that system about the same size?

Carmen: No, not really. Some had more than others as far as that goes. Where Tim Shoan is now, Ernest Rutherford had that. And he kept track of his. He always figured at the end that I owed him, or else we had extra help that would go over there to even it out.

Doug: How did that get figured out?

Carmen: Well, I don't know. He was the one that always wanted to even up! I had more to do than he had, I guess!

Doug: How close did people keep track of the exchange?

Carmen: I don't think that really they kept track. Everybody's work got done. They didn't complain too much about it—that was the only one who ever said anything to me about the difference in time. I guess he just thought that he wasn't getting his share. I know one time he figured that I owed him some more help, or money, or something, I don't know.

For Arthur, changing works was, like all human arrangements, replete with negotiation and redefinition:

Doug: Would the crew be the same for every workday?

Arthur: Not always, because maybe I wouldn't like to change with you. If I don't change with you, when it got to my place I'd hire two or three . . . maybe my brother'd come and help me to fill in.

Doug: Why wouldn't you want to change works with somebody?

Arthur: Well, you probably hooked me two or three times. I got ready to thresh; you wasn't there, you know? You may be alone. You might be old and you wouldn't be able to go back and work on the other farms. Or maybe you'd send Sleepy Martin . . .

Willie: Ha!

Arthur: Well, don't laugh about it! You send Sleepy Martin and two or three more like that, and the guy wouldn't trade with you anymore. He'd send a good man and he'd just take me off the wagon . . . take my team . . . and put me to pitchin' . . . they'd put some old guy on to load. But then when I got it loaded they'd say, "Take it to the barn and throw it off." You know? I had to do both! That's why I didn't always like that changing works. It depended where I was workin'.

Now there was one guy up the road . . . he come with a pretty good team, pretty good men. You threw him out a load of grain and he'd come out and throw it off—he'd load it into the thresher. Then he'd go right back to the field and get another load. Well, he was considerate.

But then you'd have somebody else . . . he probably couldn't build much of a load and he didn't have much of a team. You know what I mean? He might have had a long time ago, one pretty good horse and one way back. You seen that, haven't you? And now he couldn't draw only half a load. And then there'd be a problem at the thresher because he'd be back to the field before a big load got loaded. And then the fellow that was throwin' off the big load, he'd throw that load off and the other guy'd set there most of the time. It could be an old farmer or some poor farmer. . . .

Doug: Well, what about if one farmer had twenty-five acres of oats and the next guy had ten acres of oats?

Arthur: Well, we used to do it and it didn't make any difference. Just . . . [*snaps his fingers*] . . . it's gone. The time was just gone. If you went up to pay a guy he'd be mad with you.

Willie: He would.

Arthur: Oh yeah. If you was to come to pay, my father he'd shoot you!

Doug: But you say you had to hire guys sometimes.

Arthur: Well, if you didn't have enough help you had to hire more men. You didn't hire the ones you "changed works" with. You hired extra to get enough men there.

I asked Jim Fisher what it was to work together for years and years, whether people's idiosyncrasies played a role. Jim answered:

> Oh, yes, yes. There was always chuckling about someone who tried to outdo the other one, and go faster than anybody else—there was always a lead horse, you might say, in the group. Someone would be pitching faster than anybody else . . . well, that's just fun and part of human nature. There were those who had eccentricities about their horses. Their horses would always have to have grain, and in their routine, they would have to be watered first,

and then get their grain and then their hay. And there were a few particular ones that, you know, were chuckled about . . . well, everything would have to be so-and-so for his team. Then there were always the more easygoing ones . . . that's just the fun of it, of course.

Stooking

The family itself generally cut and stooked the grain. But sometimes families hired extra workers, often high school boys, to help out. The grain was cut with a reaper or "binder" pulled either by a team of horses or a small tractor, and workers stacked the cut and bound sheaths of grain into small pyramids so they would shed rain water as they dried, ripening to be threshed.

Stooking grain took practice, and while it was not exactly skilled, it had to be done correctly. Since the machine cut and bound the sheaths, the workers who stooked the grain worked, in an indirect way, to the pace of the machine:

> *Willie:* You had a carrier on the back of the binder which held the sheaves until you got about four or five, whatever you wanted. Then you tripped it and it would drop it in a little pile. You didn't scatter 'em down.
>
> *Arthur:* And then you came along and stooked them. The stookin' principle was to keep the heads of the grain off the ground to keep it from gettin' damp. If you leaned them right they'd stand up. If you didn't lean 'em good, Dad would send you back out.

> *Jim Carr:* You had the old reaper that cut it and put it into bundles. Then you had to take the bundles; it had kicked 'em down on the ground; you'd set them up; it was called "stooking it up." You put two this way, two this way, and then fill in the hole. Put six in the bottom. Two guys take their pitchforks, walk down through, and stand 'em up 'n' make stooks out of them. When it rains, see, the water'd run right off it. If they blew over, you'd go out and stack 'em all back up again.

Buck describes the competition between workers to keep up with the binder:

> My biggest pride was when the binder pulled out of the field I had set the last stook behind him. So Howard Cody came out and got me one time to stook oats. He had two men out stooking, so they two stooked the four rows, and I took four rows by myself. It was a big field. I was all away around the field with my four rows, and they wasn't two-thirds around between the two of them. But back in them days, when I was sixteen to eighteen, I would try to do as much or more than any older man he had there. As you get older you get smarter. They were the smart ones!

Fig. 53 Threshing wheat on the Ralph Greer Farm, North Collins, New York, August 1945. Photograph by Saul Libsohn.

This is a photograph of a threshing machine being transported from one farm to another in the changing works harvest.

Families and neighbors did most harvest jobs, including stooking, with no exchange of money. But sometimes these jobs were hired, usually because a farm family did not own a specific machine or tool.

Some farmers who purchased machines to work for hire, like Carmen Acres, were willing to invest borrowed money in order to try to make a return greater than their costs and interest. Other farmers who did "custom work" (as it is called) might be, as Arthur describes subsequently, the aging father with a little money and time on his hands after the farm was taken over by the sons. But it seems that nearly all farmers, on some occasion, purchased equipment and sometimes hired themselves to their neighbors. Custom work was integral to the changing works system, more common than unusual. Arthur describes custom binding:

> *Doug:* Did every farmer have a grain binder?
> *Arthur:* Well, not everybody, but I'm gonna say half.
> I say that because we used to do that. Elwood and I used to run one of them binders. We had a steel-wheeled tractor and a seven-foot binder. Seven-

foot swath. Ten, eleven acres about all you could do in a day's time, from the time you get started to the time you had to quit. You got a dollar and a quarter an acre to cut and bind it. Ten acres a day. Twelve to fifteen bucks. That was big money!

But most families timed the cutting of their grain to be ready for the arrival of the threshing crew.

Willie: You figured the regular stops with the threshin' machine. And you had your grain cut accordingly. You had to cut it three to five days ahead of the time the thresher was supposed to be to your place. You got to keep 'em off the ground because once the seed hit the ground it would start growin' again.

Mason: Yeah. After they got ripe, they'd grow. And if you got a lot of rain sometimes they'd grow in the stook; If you didn't keep 'em turned out so they'd stay dry.

Willie: And if they did that, it spoiled the rest of 'em. You couldn't pick 'em up.

Mason: Oh, you couldn't put that musty shit in the mow!

Willie: You had to keep it separated because it would catch the barn afire.

Threshing

Threshing on small farms balanced several variables. The small dairy farms generally grew ten to twenty acres of oats, which could be threshed in one or two days with a neighborhood crew typically using six or seven wagons. That meant that there had to be a thresher for about every eight farms, because all the grain in a region would ripen at the same time. Grain is more delicate than corn or hay, and the harvest had to be completed within a narrow time frame. So while any given farmer only needed a day or two of threshing, he needed to be part of a crew which would work for ten days to two weeks on the moving threshing operation. The balance between machines, available labor, and the timing of the seasons was the key to the system, and it worked well for decades.

In the early 1950s, a thresher was one of the most expensive machines a farmer might purchase. In the great expanses of the Midwest, threshermen were full-time harvesters rather than farmers who included threshing in their other jobs. In neighborhoods of small farms, the thresherman (the one who owned the thresher) was a neighbor in an entrepreneurial frame of mind.

In the North Country, the ownership of the thresher and the payment arrangements for threshing varied:

Arthur: Who owned the thresher? The fella down at the bank! Well, I mean he might be a guy that retired. Or he might be a guy that owned a farm and the two boys run the farm and the old man'd take the threshin' machine out. . . .

Willie: I think there was quite a bit of that.

Arthur: I could show you a dozen places up the road. . . . They'd own the threshin' machine. The old man'd go threshin.' This wasn't too hard a job; it wasn't baggin'!

Willie: The farmer that owned it was a little better off. I could never figure how they figured they were making much on it. They used to charge so much a bushel for threshing it. I think it was two and a half cents a bushel when I first remember it-then it went up to a nickel a bushel—eventually ten cents a bag. The theory was that it would be making money before it wore out. I didn't see it.

Since the thresher was hired, the threshing was organized to keep the machine running continuously:

Mason: You wanted at least four wagons or five . . . you kinda' planned it, you know. If you had quite a ways to draw it, sometimes you'd need five or six wagons, you know, to keep the threshin' machine going. 'Cause you didn't want that guy settin' there idle. You're gonna pay him by the hour or the bag or whatever . . . you wanted to keep him motored!

While threshing was usually organized around this combination of payment and contributed labor, some threshers worked as part of larger obligations:

Jim C.: Mr. Martin, who owned the threshing machine, never charged a cent for the use of the machine. We'd help 'em hay, ya know, whatever, and spreading manure. He just did it for the *neighbors.* But he wouldn't go to other neighbors. There was only our place and his, and down where Stewart lives. There was only four or five places where he done that. He wouldn't take jobs, you know. He could've done that work out. Sometimes there'd be twenty, thirty people call up, but the machine was gettin' old and he said he wanted it for himself. He said, "And if I take it and do yours," he says, "I'm gonna ruin it and I ain't gonna have it for myself." So, he just wouldn't do it.

The tasks surrounding threshing are documented in the following photographs and interviews. Threshing was simple work, once the workers were secured. Farmers, and possibly hired men, loaded the stooked grain onto their wagons, drove the wagons to where the thresher was set up, unloaded the grain into the thresher and drove back to the field for more. Most of this work was what Arthur called "bull work": lifting, throwing, and carrying:

. . . you go round with your wagon and pile it on. Then you draw it up to the threshin' machine, throw it in, and the straw goes out in the pile, and your oats come down in your bags.

Fig. 54 Bringing load of wheat to thresher, Betzler farm, Interlaken, New York, August 1945. Photograph by Charlotte Brooks.

Buck looked at the photo (fig. 54) and said simply:

> That is a poorly loaded load of whatever it is!

Willie explained:

> That's stooked grain. Stooked in piles.
>
> You see way in the background, how round those are? Those are stooked. But when you got done stookin' 'em like that, you took another one . . . or two of 'em, and put them on the top and spread 'em out, so that they'd bend down over the top of the others to keep the water runnin' off. That's what we used to do.
>
> *Mason:* Of course, every once in a while somebody would hire you to go set up oats for 'em, you know, like if the neighbor didn't have anybody to help 'im.

Arthur: Yeah, when the crew came, they brought the wagon, like this. But I used to work for Will Henderson. He had a good team of horses and a coon rack; there were sides on it. And I set it up for him.

It was probably as good a crop of grain as you're gonna see. Not too high. I says straight that I was gonna do it to him. Now he was seventy some years old. And maybe I would be [*clicking his teeth*], roughly, forty. But I was gonna do it to him. 'Cause he told me that he'd never had pluggin' with two people workin'. "Well," I said, "you ain't gonna have that to say again!" [He was referring to throwing grain into the thresher so fast that it plugs up. This is an example of stories farmers tell of outworking the machines.]

I set this grain all out. And I carried the bundles so they'd be just right. I set 'em on the wagon so I could get two bundles every time. I laid 'em up on there. I got 'em unloaded . . . and he gets down. He throwed his pocketbook . . . one of them long ones. You remember those long ones, Willie?

Willie: Heh . . . heh . . . heh . . .

Arthur: He give me a dollar. "Never before in my life have I ever been plugged up." He didn't know where to go with it . . . it was comin' so fast. And he was used to two men.

He was a good man too. But he'd never thrown two over bundles at a time!

Jim Carr: We'd set the thresher below the toolshed; we'd blow the straw in a pile and you ain't got very far to draw the oats to the barn. And then you take your hay baler to your pile of straw. After you threshed all day, at night, after everybody has their milking done . . . wherever you were threshin' gets the hay baler out, and you keep forking the straw right down and putting it in the baler. The bales would come out and you'd pile 'em up. Kids get up there and push it down on one another and Jesus Christ, have a hell of time, you know. I mean, it was a lotta' of work but we still had a lotta' fun, too.

Mason: You could usually do a farm in one day. It all depends on what you had. You kinda' run into a second day. And a lot of times if you got done a little early, the guy that owned the thresher, he'd move into the next place and set up for the next morning. And, of course, it all depended if you had enough help!

Willie: Now, these belts (figs. 56, 59), every time they stopped usin' it at night them belts was all taken off and stored inside the machine. Either that or in the guy's barn. Then he come back out in the morning and that's the first thing he

Fig. 55 Hired man standing in straw stack, Betzler farm, Interlaken, New York, August 1945. Photograph by Charlotte Brooks.

Fig. 56 Threshing barley and oats at Wheeler farm, Dryden, New York, August, September 1945. Photograph by Charlotte Brooks.

did, put them belts back on. They'd start slippin'. You take 'em off and let 'em relax overnight and put 'em back on. Just like puttin' a new belt back on.

Willie: You forked the shocks on by hand just the same as they are forking it into the threshing machine. See here, one sheaf at a time (fig. 57). There are knives here where it first goes into the threshing machine—the knives chop that all up and it goes through a beater like.

It goes in on a beater, and it breaks the straw and everything up. The grain—this is oats, you can tell—goes over a screen, and the grain falls down through the screen and the straw goes on out where there is a blower that blows your straw and your chaff. That is why you see so much dust in the air here (fig. 58)—blowing that chaff out that it beats off. You would get more of an idea of how it works if you took one apart. All through the inside of the thresher there are screens—your grain goes through and goes into the bot-

Fig. 57 Threshing wheat on the Shaughnessy farm, Warsaw, New York, September 1945. Farm hands toss sheathed wheat into threshing machine. Photograph by Sol Libsohn.

tom, and on the other side, where you see these pipes here come up—there is an elevator—it elevates that grain up. Then there is another elevator—screw elevator in this one, that brings the grain out where he is bagging it.

This elevator recycles it; brings it back up and recycles it if it isn't all clean. The elevator picks it right off the bottom and it goes through the screen—it takes it right up and recycles it.

Arthur: But you got a good photo there; he's throwin' it off the wagon into the threshin' machine (fig. 57).

Doug: They don't have to untie 'em?

Arthur: No, no.

Willie: They've got knives in there . . .

Arthur: At one time they used to untie 'em. Remember that?

Willie: Yeah, you used to have to cut 'em.

What do you see wrong in that picture?

We always used a sheaf fork when we was handlin' the sheaves. That's a three pronger there and a sheaf fork's a two pronger.

Fig. 58 Threshing on the Norman McIntyre farm, Perry, New York, August 1945. Farm hand on top of straw stack in barnyard stacks the straw as it comes out of the threshing machine blower. Photograph by Sol Libsohn.

Arthur: Yeah. Maybe. We never used a sheaf fork.

Willie: Oh, they were handier; didn't stick in the sheaf or anything. Toss 'em and keep goin'.

Arthur: Yeah. Two fork . . . two tines. They was a little closer together than that one.

Now, this is something amazing to me. Here's a guy . . . two guys . . . one's got a cap on and one's got a hat. Here's a fellow workin' with his hat; the other fellow with a cap. But some people would wear a cap. Some people would wear a hat. Right?

Jim Carr: Boy they'd do 'er! See there it shows. There's your pan where he's throwin' it in. He's standin' up on the load. You'd throw your first bundle in—well, it depended on how big a tractor you had. If you had a big enough tractor you could lap them a half and not plug it. Or some of 'em you'd have to space 'em a little bit. But if your old tractor had lots of beef, you could lap those bundles a half and put a lot more through. But if you put it through too fast it don't clean it as good, so you got to watch.

Fig. 59 Threshing wheat on the Shaughnessy farm at Warsaw, New York, August 1945. Hart-Parr tractor used as power takeoff for the threshing machine. Photograph by Sol Libsohn.

There's string on the stook and when it goes in through there's a knife that'll cut the string. It just gets chopped right up. But you see your string is smaller than bailer string. And your stook is probably only like this big around, so you know, it ain't that much.

Jim Carr: This is where it's blowin' out into the straw pile (figure 58). He's probably up there levelin' it off or somethin'. Because after your pile got so high your spout couldn't clear it; someone'd say: "Run up there and knock that down." And that son of a bitch'd be a-blowin' black shit out of there. Heh . . . heh . . . dirty. . . . Oh, Jesus.

Arthur: And that was a black job.
Doug: A black job?
Arthur: Dirt. Dust.

Fig. 60 Threshing on the Norman McIntyre farm, Perry, New York, August 1945. Farm hand tosses sheaves into the thresher. Photograph by Sol Libsohn.

Willie: That's runnin' the threshin' machine (fig. 59). It ran off the tractor. You kept the tractor far away because of sparks comin' from the exhaust. Like if it would bog down and you'd crack the governor on the tractor, usually you have sparks comin' out of the exhaust. So you kept your tractor far enough away. You're probably a good . . . forty feet away from the machine with this belt.

He's probably the guy that owns the outfit. He's just standin' there watchin' it. That's what the owner used to do. If he wasn't watchin' it he was feedin' it. Keepin' his eye on everything. Yeah. Feedin' it like this here. Because one thing they kept a close watch on was fire. See, this outfit is set up out in the field for their threshin' because you see the stubbles right around the tractor. So when they thresh out there then they've got to take the sacks of grain back to the barn. They'd load 'em on a wagon and haul 'em back to the barn.

Arthur: The thresher ran off the tractor, and you wanted a big one, too.

Willie: That one right there shows you. Yeah.

Fig. 61 Threshing barley and oats, Wheeler farm, Dryden, New York, August 1945. Photograph by Charlotte Brooks.

Mason: That looks likes threshin' loose stuff though don't it (fig. 60)? It's probably what was on the ground. . . . Most of your guys that bring a threshin' machine around didn't like to put loose stuff through it. Afraid of gettin' stones.

Willie: Oh, it'd mess up a machine in a hurry.

Mason: Yes. Stone'd go through, mister, and the spark'd go right up into the mow.

Willie: Fire.

Mason: Yes. You bet!

Willie: Cause a fire in that dust. Quite a few barns burning from that.

Mason: That dust would explode by itself if you lit a match.

Willie: Now, the old machines that they came out with first, you couldn't throw 'em in like that. You had a guy standin' here cuttin' the string.

The new style, they had knives in there. Just like the sickle might on your mowin' machine. And those knives kept workin'. They'd cut the twines. And it'd cut it in pieces 'fore it went by there.

Mason: Mister, there's a job right there, carryin' bags.

Willie: Oh, I hated that.

Mason: Oh, geez, I carried them sons of bitches up stairs and everything else. Some of 'em had the granary upstairs. . . . And then they had a chute that you spilled the bags though, you know. Which was handier but it wasn't so handy when you had to carry them oats up there.

Willie: Some of 'em were three-bushel bags; two, two and a half at least. More than a hundred pounds, maybe a hundred twenty.

Mason: Oh boy. They was heavy enough, I'll tell you. Especially at night; they got heavier!

Arthur: You would have dumped one side and took that bag off; let the other bag fill. Put another bag on there. By the time you got over there that one would be full, if you had a good outfit. Dependin' on who you had up on top of the wagon and how many oats was on the granny . . . heh?

Doug: So the combine (fig. 62) replaced the thresher?

Arthur: Not mine it didn't! A combine never got as good a oats as the stand-up oats you got with the thresher.

Doug: What do you mean?

Arthur: Well, when you cut the oats and stooked 'em up for ten days you got a better quality. The difference between oats out of this and the stood-up oats'd be so much a bushel, two or three pounds. Your thresher would take the oats out of the straw better. That the way you've heard it, Willie?

Willie: Yes. There was quite a bit of green passed right straight through a combine—all that shakin' and everything.

Arthur: A lot of little oats was gone.

Doug: So this was a step down in quality.

Arthur: Well, as far as I was concerned. I'm a Scotsman.

Arthur: Sunday if you got nice weather—Sunday you combined (fig. 62).

Willie: All you had would be one guy to stand on the combine baggin'.

Arthur: Elwood had one with a bin on 'em; no guy needs to be standin' there

Fig. 62 Combine on barley field, New York State, August 1945. Photograph by Charlotte Brooks.

baggin'. He'd pull up to your wagon and you would hold the bag. He'd open an auger and he's watchin' it . . . sometimes you'd get you too full a bags didn't ya?

Doug: So, all this cooperative labor, it all went out the window?

Arthur: That was gone. Yep.

Doug: What about that? What was that like? Did you miss, it or was it a relief?

Arthur: Well, you see, the trouble was the war come along and there was a lot more jobs. More jobs in other places. . . . All this stuff had to come.

Combines came and left the North Country, for all practical purposes, within five years. But combines had been in use for nearly a hundred years, used in different settings and on a different scale of farming. Canine tells us that the first combines were made in the 1850s, drawn by horses and powered by steam engines.[23] The invention of these machines was stimulated by the population influx into California after the Gold Rush and by the subsequent development of wheat farming in the dry and fertile Central Valley. By the 1880s, steam-powered, self-propelled combines, which cost more than five thousand dollars, cut forty-foot swaths, harvesting over ninety acres a day. They were designed for miles-long fields and a dry harvest. They were part of the early transformation of farming from a family affair to a corporate enterprise. These machines had no use in the one-hundred-fifty-acre farms of the northern dairy regions, where most farmers grew ten to twenty acres of wheat.

Threshing, as we have seen, worked well in the North Country (and in all northern dairy neighborhoods), but threshing required crews of eight to twelve men and four to six wagons.

So the small combine had a place in the evolution of North Country farming, but the history is rather complex. Farmers began using combines partly because of the decline in the number farm workers during and after World War II, and because of the farm consolidation during this era. But probably more important was increased corn production, which I will discuss in the next section. In the meantime, some farmers, who had been part of changing works threshing groups, bought combines and did custom combining for their neighbors. Jim Fisher describes this process in the local neighborhood:

> Combines had come in earlier in the West, but they didn't come in evenly. Carmen came here from Canada in the forties, and he had run combines in Canada as a boy. He got the first combine over here, and replaced Jess Walker's threshing machine, the handling of grain, stooking it all. The fields looked beautiful, all stooked, but it takes a lot of back work on a hot day to stook acres and acres of grain. But when you get it done, it's a picture! But then you run the risk of going into two or three weeks of wet weather, and maybe seeing the oats growing right in the stooks and losing a lot of goodness of it. So, the combine was welcomed.

Carmen describes the transition:

> After I had a combine, Beckstedt bought one—two years after. They went around combining. And there were a couple over on the river road—self-propelled jobs. Then Beckstedt sold theirs and the one the fellow on the river road owned was wore out. And then, I more or less quit doing it. The machine was getting old, and I'd done it for four or five years—I was getting sick of it! I figured if I quit using it, it would do for me as long as I wanted it. If I kept going for others, I'd have to buy another one. People quit growing so many oats, too. They were working into more corn, I think, and more silos, and more forage because it was easier—it came down to one crop more or less. You'd get more cows and have to feed as many cows as I'm keeping. So, in short, you can get a lot more tonnage off an acre of corn than hay.

Figs. 63–68 Combining oats, the Acres farm, 1978: the end of an era in the North Country. Photographs by the author.

Jim Carr discusses the impact of labor shortage on the transition from threshing to combines:

> *Doug:* Were there a lot of years after the threshing stopped, when people used combines? My sense up here was that people stopped growing oats . . .
>
> *Jim C.:* Well, In later years that's why we bought the combine, because you couldn't thresh. . . . There wasn't anybody that had a threshing machine or nobody did it anymore. So we bought that and did it ourselves.
>
> The threshing crews were neighborhood crews, but in later years people moved away and so we just bought it and did it ourselves.

Because threshing crews disappeared, farmers who grew oats and other small grains came to depend on a declining number of deteriorating combines in the neighborhoods. Jim Carr adds his account to Carmen's:

> When we had the combine, we could have been goin' all the time if we'd wanted to. But we did ours—oh, If a neighbor was in a pinch we'd go do it for 'em. But as far as goin' out and doin' custom work, we said, "To hell with it." By the time you furnish the machine, and you had to furnish the guy to run it, and the tractor to draw it. And somebody to drive the tractor. And if they didn't have anybody to bag it, you had to send them, too. So we said, "It ain't worth it." Put it in the barn and save it for yourself. You know? That's the way we looked at it. So that is the reason people stopped growin' oats; it got too hard to get 'em off.

Still, a few farmers continued to grow and combine oats for a few years after the SONJ photographs were made. The early years I lived in the North Country I would occasionally hear a combine on a back field. It was an enormous machine and an extraordinary sound, rather like a jet airplane on the runway. This particular year I had watched a field of oats ripen and expected Carmen to chop them for fodder. But one fall day the combine was out, as it happened, for the last time. I walked to the back field and photographed the great machine and Carmen's obvious pleasure in running it. He invited me for a ride: the swan song for an old beast.

The Corn Revolution

Within a farmer's lifetime, the role of corn on family farms was transformed. More than anything else, the changes surrounding corn made the modern dairy farmer a mechanized capitalist rather than a participant in a neighborhood of near equals. The SONJ photographs capture the middle stage in this revolution and introduce the technology still in use.

The transformation of corn was due to several developments, inventions, and changes in cultural practices, which I will briefly describe.

First, corn itself changed during the twentieth century as new genetic understanding led to modern hybridization.[24] As a result, corn production (and the price of corn seed) increased several times over between 1920 and 1950. By the SONJ era nearly all farmers were using hybridized rather than open-pollinated corn, and the increased yield of this corn itself argued for bigger and bigger plantings.

The second important change was the replacement of mechanical control of weeds by herbicides. Because farmers had to weed corn before herbicides were developed, there was a natural limit to how much they could grow. Cultivation was once done with hand tools; then, in the nineteenth century, horse-drawn mechanical cultivators came into use. These were small machines that scratched at the weeds, uprooting them and covering them with soil but leaving the corn intact. They were updated for use with the small and agile first-generation tractors of the SONJ era.

Cultivating filled in all the extra working hours during the first months of the summer, and it limited the acreage of corn which could reasonably be grown on a family farm. Carmen describes how his family cultivated a ten-to-fifteen-acre field:

Fig. 69 Cornstalks being artificially pollinated, Cornell University agricultural experimental station, Ithaca, New York, September 1945. Photograph by Charlotte Brooks.

You would cut hay every day in the morning, and then wait for it to dry, then run out and cultivate a couple acres of corn with the horses. You'd work it in whenever you could, or after a rain some days, when it wasn't too muddy, you go down there and cultivate until dinnertime. The next day you start off from there, and whatever time you had to fill between the jobs. Then after you cut higher, you let the ground roll in over the top of the weeds onto the corn. We used to have little pans on the cultivator which would run along the edge and keep the dirt from rolling up into the corn, because otherwise it would kill the plants. Then after the corn got up to a foot higher, you take the pans off and let the dirt roll over in between the corn, because the stalks would carry it then. But I couldn't see cultivating a hundred acres like that!

Farmers stopped cultivating when herbicides were invented during World War II (perhaps ironically, as a result of germ warfare research). Herbicides (and, later, pesticides and artificial fertilizers) replaced labor with a new expense. Farmers had been able to cultivate their own fields, but they needed to pay for the application of herbicide, which came to be called spraying. Killing weeds with herbicides removed the natural limit on the amount of corn which could reasonably be grown, and, because corn produces more feed per acre than other crops, an increase in potential corn production increased the potential size of the dairy herds. Finally, adopting herbicide use moved farmers toward a world in which chemical solutions were seen as reasonable. When the farmer accepted this world, it was easier to replace crop rotation with artificial fertilizer and to use pesticides to kill insect populations which increased in the monoculture of corn.

All aspects of corn work were transformed in this process. Until the 1930s, corn was planted and harvested by hand. Farmers walked through the fields "broadcasting" seed in a practiced swing of their arms. They cultivated it by hand during the hot months of the summer. They picked and husked it by hand in the fall and probably handled it several times in the process of livestock feeding and in discarding parts of the plant:

Willie: I can remember when, after they got their chores and things done at night, the guys would all get together and go cut and peel the corn. That's the way they got their corn.

Arthur: They'd cut it by hand, with a sickle.

Willie: Yeah, four or five guys going down through the field cutting. And two, maybe three guys tying. We carried a bunch of twine already cut. Pulled twine every time you'd get a sheaf.

Arthur: There's something' that we used to do.

Farmers remembered the hard work, but they also remembered how neighbors used the event as an excuse for a harvest party:

Fig. 70 Joseph Kujawa, West Webster, New York, farmer, gathering sweet corn, August 1945. Photograph by Sol Libsohn.

Mason: Used to have "huskin' bees." Remember? They'd have a huskin' bee; everybody'd come and help. All the neighbors come in and helped and have a big jug of "mountain dew."

Willie: Mountain dew or hard cider.

Mason: And then, of course, when you got that done you had a fiddle and a piano. You'd have a little dance. Little party. This is in late fall. Be in the winter partly, you know. They'd go out and pick the ears off and bring 'em in. Every time they got a red ear, why, you'd get to kiss one of the women.

Doug: So when you went to huskin' bees you went for the . . .

Mason: Women. And helpin'. You know.

Jim Carr tells a similar story:

They'd have a whole bunch of corn stooked up and you'd draw it to the barn. And then all the neighbors'd come and husk it. It was just about like threshin.' They'd have a big meal and, you know, you'd go. The whole family. Kids and everybody. And you'd probably get started at seven–eight o'-

clock at night and probably husk for two or three hours, and then have a meal and then, you know, have a party, maybe a dance.

Doug: What's this thing about who finds the "red cob"? Did you ever hear of that?

Jim: Yeah. I guess if you find a real red ear you're supposed to kiss a woman that's near ya or some friggin' thing. I don't know. But that's the way you'd get your corn husked. You have twenty–thirty people come, you get a lot of corn husked.

Corn huskers used a tool called a "husking peg." Willie remembers:

> It laid across your hand and it had leather straps that would hold it to your fingers. . . . You grabbed a husk and pulled it off. You didn't pick the ear of corn; you husked it right on the stalk so you didn't loose the husk. You skinned it back and broke the ear off. You leave the husk right on the stalk. That was one of the best eatin' parts for the cow. They'd hunt through your corn pile to find that first.
>
> You threw the cob on a pile or a wagon'd be comin' through. We threw it on a pile most of the time and . . . then the guys come through and they used forks to throw it in the wagon.

The revolution in the corn harvest involved the development of the stationary "chopper/blower," which was in use by the 1930s in the North Country. These machines spun knives at high speed, chopping stalks, ears, husks, and cobs into small pieces. The pieces were then blown into a long pipe to the top and then into the silo. As this green, high-moisture corn fermented, it made silage. If it was too dry, or if the chopped corn contained air pockets, it molded and spoiled.

The use of the chopper and silo made it possible for the first time to feed cows stalks and cobs, the previously less palatable parts of the corn plant. As a result, the amount of feed realized from a corn plant increased. It also became necessary to build and maintain silos. The first were three- or four-story wooden or cement buildings, and later ones were great and complex towers costing up to a hundred thousand dollars. To work correctly, silos needed to be nearly airtight. Gases formed as a natural part of the fermentation process, which added a new risk to farming. Jim Carr remembers a silo accident:

> There was a bad accident in a silo, up to the Teal farm. Up on Route 37, where Paul Dalton lives. They had a Harvester silo; it was full. The young fella went up and the corn was up near the hole where he was gonna slide the metal roof. You could slide it open, and shut it after you were done with the silo. Well, he had a fork, a pitchfork. He was pushin' the corn away from the hole. Well, he loses the fork. So what's he do? He jumps in the silo to get it. He jumps in the silo and the pile takes him. In those Harvester silos

you're not supposed to get in 'em because there's fumes. And he couldn't get out; his father heard him hollerin'. So he went up; tried to get him out. He jumped in too, and the gas killed 'em both. The kid was about fifteen–sixteen years old. Well, he'd just come back from the Gouverneur fair; the Lisbon band had been in to Gouverneur fair that day. That was in—let's see, it was in '67 or '68 'cause Cheryl was a baby. Doris and I went to the fair that day and we'd just got home and went down and picked up Cheryl, and came home and my father told me that Teal and the boy both were killed in the silo. Then they called the rescue squad and the sheriff. See there's a big fan. If you're gonna work around them or do anything with those silos, you gotta run this fan for so many hours before you can go in there. The fan wasn't on and one of the sheriffs says to Russ Good, "Go right up and get 'em out." Well, Louie Brander, that lives up to Lisbon, he's had them silos ever since they came out. And they called him and he come right down. And he told 'em, says, "Turn the fan on. Don't go in the silo for two hours." "Well, we got to get 'em out." He said, "They're dead." Said, "If anybody goes down in the silo you're gonna be in the same friggin' shit."

See, it's glass inside, and once you jumped in there ain't nothin' to get ahold of to get back out. They have a glass liner in 'em, like a thermos bottle. The kid figured his pitchfork would screw up the unloader when it come to the bottom. Well, you don't think of the consequences. He lost his fork so he just jumped in to get it. And once he was in there it didn't take very long for the fumes to get him.

This tragedy was retold in many versions. When the farms mechanized, farmers began using dangerous machinery. Many parts of the machines spin at high speed in unprotected shafts. Tractors are bigger but not necessarily more stable. As the weight of the machines increases, the consequences from a small mistake loom more ominously.

A tractor supplied power to the stationary corn chopper through a removable belt (fig. 75). A farmer could do all the steps involved with silo filling, as the job was called, but it made more sense for crews with four or five wagons to bring the previously cut and bound corn from the field to the chopper, where the chopper ran continually. Thus, silo filling was a natural use for changing works crews, and, indeed, for at least five decades, North Country neighbors filled silos together.

Farmers first cut and bundled corn stalks, which were left to dry on the field. Crews of two or three, each with a wagon, picked up the corn bundles and threw them on a wagon, which was driven to the silo. In the early stages of this process, draft horses pulled the wagons, following commands of the farmer working alongside. When tractors replaced horses, usually a boy, too young to work in the field, drove the tractor slowly through the fields and to the silo (fig. 73). Other workers picked the six-to-ten-foot corn stalks off the wagon and fed them through the blower/chopper. One worker, usually too old for the field work, worked inside the silo, directing the stream of silage evenly throughout the structure and tramping it down so it would not develop air pockets and mold. The division of labor, rudimentary as it was, was based on age and physical strength.

Changing Works in the Corn Harvest

The corn harvest was longer than the grain harvest—often up to six weeks, and later, so that much of the work was done in the cold and wet part of the fall:

> *Arthur:* Who started first? There was always a Baxter.
>
> *Doug:* A what?
>
> *Arthur:* The guy that had cut it down and wanted his put it in the silo—you had to go help 'im. It wasn't a place to argue. Not very often. Once in awhile somebody'd get a little upset. But usually you knew Old Frank Baxter's gonna put his in first. And then he'd be good about it. Everybody'd go help him. And he had two boys . . . three boys, and himself. He'd run the blower. He wouldn't charge you too much. And he'd send some good boys and a team. Good ones. I ain't talkin' about the poor ones, I mean the best. So we would always fill his first. But he'd bring the blower and he didn't cost half as much as havin' 'em do it with another guy. Wouldn't be long there. And if you was gettin' behind, he'd send the boys up at night when they'd get through milkin'. Sent one guy up with a team, maybe four o'clock . . . he sent him to go get the team and he'd cut corn maybe 'til . . . if it was a moonlit night maybe he'd cut a few rows after moonlight. You don't believe that, do ya?
>
> *Willie:* Oh yes.
>
> *Arthur:* Heh . . . heh . . . We used to cut corn by moonlight!

The SONJ photographers recorded most of the steps of the corn harvest. They also photographed the machine that made the changing works crews obsolete: the mobile chopper, which cut and chopped the corn as it was pulled through the field. The mobile chopper was one of the most important contributions to the transformation of the corn harvest from a communal, neighborhood affair to an individually controlled, mechanized job.

The development of mobile choppers (which, for no reason, but to great effect, coincided with advent of commercial herbicides), led farmers to increase corn from five to ten acres per farm (about 10 percent of a typical farmer's crop during the SONJ era) to currently more than one hundred acres (50 percent of a moderate-sized contemporary farm). The revolution in corn increased the tonnage of feed per

Fig. 71 New England farmer's wife binding corn stalks, North Hatfield, Massachusetts, October 1947. Photograph by Gordon Parks.

acre, allowing for bigger cow herds and income from milk checks. (Farmers typically call their monthly payments for milk sold "milk checks.") Of course, it also increased farmers' expenses; it led to the loss of their working contact with their neighbors and turned the farmer from ecologically reasonable methods.

In the following, my intent is to show not only how the harvest was done, for that is quite simple and, in fact, already described. Rather, I would like to hint at what it felt like to be a part of these harvests and to suggest how farmers defined the changes of which they were themselves a part.

Fig. 72 Cornfields between Medina and Lockport, New York, August 1946. Photograph by Gordon Parks.

Arthur: Where's that woman? (See fig. 71.) She's stookin'. She's probably pickin' some husks or ears off that stook, but it's all stooked up there. In the winter, you'd take the horse out and put the log chain around it and tow it up to the barn and put some in the manger.

You feed the whole bundle. That was generally bundled. It could be loose. But you'd take the whole thing up to the barn . . . you'd throw that in to the cows and then the next day you'd take out the stalks. You ever do that?

Fig. 73 Gathering corn for ensilage, Brooklea Farm, Kanona, New York, September 1945. Photograph by Charlotte Brooks.

Willie: I hated that.

Arthur: Oh, so did I. They wouldn't eat the stalks.

Now, I just remember when they did that. Now, I was born in '16 and when we was . . . oh, probably . . . when it got '30 . . . 1930 they was doin' it this way.

Willie: I can remember that we did that.

Doug: You said that the animals wouldn't eat the stalks.

Arthur: Heh. . . . Burn 'em. Didn't we?

Willie: A lot of 'em burned 'em. A lot of 'em chopped 'em up and put 'em back on the field.

Arthur: Yeah, but we didn't have no way of choppin' 'em at our house, except by hand.

Willie: Yeah, I've done that too.

And then there's places where they would cut the corn (well, they still do) a foot and a half, two foot high, and leave that on the bottom. Then you pull that last two feet out on the ground to make fertilizer.

Willie: There were some times you didn't get it picked up until after the snow had hit it. But you wanted to get it before the first snow. You wanted to get it before it froze into the ground or you didn't get it.

Mason: I've been to huskin' bees when the snow was on the ground.

Doug: Why'd they want dry corn?

Mason: Oh, they always had pigs; or they'd grind it up for cattle.

Willie: 'Bout everyone had shellers. They'd shell it for their chickens.

Mason: This was a different brand of corn. We grew silage corn and this was huskin' corn. It growed faster and dries faster. Now If you couldn't get your silage corn all in the silo, why, you'd blow it in a pile . . . and then feed it first, you know. But this was a different type of corn.

Jim Fisher: We usually grew about twelve to fifteen or sixteen acres of corn (fig. 73); two wooden silos—one square and one round, and we never had over eighteen acres of corn. That seemed to be a reasonable amount of corn to have to pitch out by hand in the winter, as that was before silo unloaders, and we milked thirty cows at that time.

It would take probably three days for each farm, depending on the weather. If the weather was good and you weren't slugging through mud. Of course it was cut ahead of time; not too far ahead, because you didn't want it too dry. If, for instance you got the field cut with the horses and the corn binder and it lay—and you got a hard frost and then it warmed with the corn on the ground, the leaves would all dry out. There would hardly be enough moisture after a few days to keep the corn, so you would have to connect up a garden hose and have that running into the blower as you unloaded the corn. That made more moisture; otherwise, in these old wooden silos, you would hardly have enough moisture to keep it. It would mold if it didn't have the right moisture content.

Willie: You didn't have any time to bullshit! Especially on those corn deals.

Mason: When I was about sixteen or seventeen I went over to Delbert

Harper's to pitch corn. The cornfield started right there and went back, you know? And he had corn at least as high as this ceiling. And bundles that big [*spreads arms*].

I think I was about seventeen and Jesus it was hot that day. One of them days in the fall when the old sun was beatin' right down. There was two of us pitchin' and three wagons. Two pitchin' and three wagons.

Willie: That's a killer.

Mason: Guy Moore and me was pitchin'. Well, we filled that goddamned silo full and about four o'clock we was all done. We had 'er full. Yes sir! And me and Guy come up and, of course, your clothes was just soppin' wet, the sweat run right off ya. And, of course, I had a pack of cigarettes and book matches. And I pulled out them book matches and they were just solid; the sulfur had just dripped off. Delbert Harper smoked a pipe and he had them farmer's matches. So, Delbert Harper come around there—oh, Jesus this made me mad—I said to him:

"You got a match, Delbert?" He says, "What do you want a match for?" "I want to smoke." He says, "You're too young to be smokin." "Well," I said, "I ain't too young to pitch your goddamned corn though, am I?"

Never again. Never helped him again! And I went home and I says to the old man, "If that fella ever wants to 'change works' again—don't send me; you go."

"Why?" he said, "What happened?" Of course, I didn't dare tell 'em about the smokin', you know. 'Cause he didn't know I smoked. But I told 'em about pitchin' corn. Mister, I never put in such a day in my life, I'll tell you. Along about three o'clock you pick up a bundle of corn on the fork and the tassles'd drag right on the ground. Honest to Christ, mister, heavy!

Guy Moore says, "Mister, you ever see me back here again, you bring the shotgun and shoot me right between the eyes!"

Jim Carr: We probably cut it in the morning, picked it up in the afternoon . . . it don't make no difference with the corn. As long as it's ripe, you just get it off. But say, those bundles are heavy! You get in corn 12–14 foot high and most people who loaded it used a pitchfork. It's got a string around it too. When your corn harvester cuts it, your corn goes in standin' right up. There's a pan that runs right down along the ground, cuts the stalks off. It runs the corn right up in the machine, and when it gets so much in there it'll trip and tie it around the middle, and kick it out the back. Then it's just layin' there in the bundle. Most people use a pitchfork to throw it on the wagon.

You had a fairly long day, but by the time everybody got their milkin' done in the morning, you'd probably get started by eight o'clock. And then by four or so you'd have to quit because everybody had to go home and do their night chores.

Some guy'd have fifteen acres and maybe some guy'd only have five or six. It'd depend. But everybody went and helped until everybody was done, you know. Nobody kept track; everybody helped everybody else.

Doug: Is there as much skill in this corn chopping? Or was it just bull work?

Arthur: Bull. That's just what it was.

Doug: How about stackin' it on the wagon?

Arthur: Well, if you had a good pitcher you didn't have to stack it much. Maybe you'd have to straighten it out a little. Unless you wanted to bull.

You wanted to get this done early September, so you wouldn't work in the mud. Today, they don't know enough to get done, do they?

Jim Fisher: Murray drove our tractor—that would be our first John Deere tractor—when he was just a little boy, because he was very mechanically minded and he drove the tractor on the hay wagon.

This corn has been bound in the fields and then these helpers are loading the bundled corn (fig. 73). You pick them up either by a fork or by hand and throw them on the wagon and take them to the silo where the ensilage cutter is set up.

You start at the front of the wagon and throw off bundles onto the blower. Yes this photograph (fig. 75) is the blower and corn chopper at the silo. You start at the front of the wagon and start pitching off bundles, and get down to the rack, so that you can turn around, and then you are in business. After you have taken off probably a quarter of a load, you move the tractor and wagon ahead so that you don't have to walk very far with each bundle. They throw off and you can unload a load in probably ten minutes or fifteen minutes. The tractor is belted up to the corn chopper.

Before there were tractors, of course, it was the same operation but with the horses. And that works well too because it actually saved a person. One man got on the wagon and wrapped your lines around the standard after you started, and got your horses headed straight—you could just chirp at the horses and they would move up a step and "Whoa!" and they would stop, and you would work at the corn yourself. On the loading process usually there was just one man driving the tractor on the wagon and two loading in the field. But then there were usually three farmers with their teams there, so you would have three wagons.

You took your team and you went to the neighbors, and you took your hired man too, so that you would have the two helpers. And then you would go to your next neighbor. Probably twelve people total—because you would have three or four people with wagons. When it came time for dinner, you un-

Fig. 74 Dump truck and ensilage blower, A. E. Scudder farm, Painted Post, New York, September 1946. Photograph by Charlotte Brooks.

hooked your horses from the wagon and tied them in the barn—you just turned them around and gave them some hay. Took off their bridles and put on their halters, so they could eat without the bits in their mouth, and they ate hay and a little grain while you were in having a great feed at the big table.

Buck: Now, my uncle, at silo filling time, he was always a little bit on the tight side I guess you would call it. So when it came time to pitch corn he would open his binder up all the way so she would put out great big bundles. He got to where he couldn't get anyone to pitch corn because the bundles were so heavy. He figured he was getting more out of the men!

Willie: They would leave a young one just idle along like that with the tractor (fig. 73), and when they get to the end of the field, they would get on it and turn it. I have seen them do that many times.

Fig. 75 "Making corn," the operation of chopping and blowing the corn for ensilage into the silo, Brooklea Farm, Kanona, New York, September 1945. Photograph by Charlotte Brooks.

He had to know where he was going. Your tractors all have got governors and things on them. Set them how fast they want them to move. The boy just steered it down the field.

As soon as they are big enough to do anything—usually they are out there wanting to do it. This is probably one of the first jobs: that and feeding calves. They always had their favorite calf!

Arthur: Well, this looks to me as if he was short crew. We used to work—really to do it—put two men on a wagon—one to throw—one standin' there to keep it from gettin' plugged. Because the stalk would catch inside and it'd stop. You generally had about three bundles on top so you'd had to push the big lever that was right there. That right?

Willie: Yep.

Arthur: What you wanted to do was to keep them big stalks a-going' in there. Maybe one might be crowdin' you a little so you'd push that back a little. I've plugged lots of them tractors. Oh, lots of 'em! I'd load my loads and I could plug ya the last load before dinner. Always quit at quarter to five!

Doug: You'd plug the pipe there?

Arthur: Right. Yeah. Plug the pipe. The belt to the tractor would slide and they'd have to shut 'er off. We'd pull the end off or pound with something.

A guy brought that chopper and a tractor. There wasn't a tractor big enough to run that for a span of . . . oh, I'd say fifteen, twenty years. I'm talking the twenties, thirties. They were runnin' 'em with a five-horse gasoline Associated engine. Remember them? They'd go "putt . . . putt . . . putt . . ."?

Willie: Stationary engine.

Arthur: Cut the bundles, put in four or five stalks . . .

Can't talk . . .

Doug: You're talking good.

Arthur: Too many years.

Doug: Why is that belt so long?

Mason: The longer the better. They don't slip. And see he's got it crossed, you know, so that it runs it the right way. If you just run a straight belt it'd be runnin' the wrong way. Runnin' 'er backwards. Most of them run with a crossed belt. I've seen the wind blowin' though and you'd have to put a crow bar in the center so it wouldn't blow it off the pulley.

Willie: Many a times . . .

Mason: . . . the belt would get to floppin', you know.

Willie: The old wind would get to floppin' the belt up and down and she'd run 'er right off the pulley.

Mason: Yes sir. . . . And some of them old blowers used to blow up. Remember them old Blizzards? I think what happened is something caught or a blade come loose.

Willie: Yup. The blades . . . they wouldn't keep 'em tight. They come right off.

Willie: They used to line the silo with tar paper and—I can't think of the name of it now (fig. 76). Similar to tar paper—it wasn't suppose to give off any taste like tar paper did. But most of your old wooden silos, if you done

Fig. 76 Farmer controlling blower inside the silo, Brooklea Farm, Kanona, New York. September 1945. Photograph by Charlotte Brooks.

them up too good on the inside, they would leak and the silage would get dried out. If a guy didn't get his staves turned up tight enough, you took a chance on losing it, or it tipping. The ensilage would keep it wet just like a barrel—keep it swelled up.

Doug: What do you mean, "lose your silo."

Willie: A loose silo will fall—I have seen them go over. You know where Joe Mitchell used to live over here? His went over. Too loose. When it swelled up, it was like a barrel. The staves are tight from moisture. He had one that dried out. He didn't get very much corn in that winter. He set it back up again.

Jim Fisher: This shows how you had a man in the silo, with a long rope on the distributor pipe that hung from the top of the silo, connected to the pipe that went up on the outside. It came up on a curved arch and you hung a distributor pipe on it with a long rope, so that you could guide it around inside the silo. You carried it around and around so that there was a stream of corn, keeping the silage smooth all the way over the silo. Sometimes there was a second man in the silo just to tramp it. Often you would have a teenager doing that. It wouldn't be dusty, but it was a cold job. On a cold day with that wind coming in with the corn and blowing through! If the sun wasn't shining, and it was cold, cloudy day in the forties or fifties, that was a cold, hard job!

This photo is taken from inside the barn; there would be doors that you close as it goes up. There is a whole line of wooden doors that are set in place as you fill the silo up. There would be about thirty inches around the doors so there would be a path right up through. About every thirty inches there would be a metal cross member going across, that the doors would sit in, and then once the corn presses against them, that holds the door shut. Then as the silo is fed, you just take your doors out as you shovel the corn out. You climb up in there then you just shovel it out; probably shovel down a layer six or eight inches deep, depending how many you are feeding.

You took out your door and just threw the corn down, and that would probably go onto a barn floor, or the floor of a silo room, depending. Most of these silos connected to the barn, with a small silo room that would hold a wheelbarrow. They would probably be eight by ten feet. You would throw out this pile of silage and distribute it around to the cows with a wheelbarrow. You did this twice a day; you always got them fresh silage.

As I recall, we fed silage night and morning, so quite often, when you started milking, or before you started milking, you might feed silage, and then you would put the grain on top of it. They would be eating and comfortable while you were doing the milking. You couldn't feed the silage too long before you milked, or they might get a gassy, silage smell on the milk, so that was something to think about.

Well, this is an interesting picture. I had even forgotten that whole—

Emily: —I did too—

Jim: —scene of the silo, because from the time I was a little boy I would hurry home from school and climb up in the silo to be up there to help tramp the silage. If it was on a Saturday you would be up there with your grandfather probably—usually it was an older man that was assigned to the silo business because he just directed the corn around with the pipe. Or sometimes two teenagers would do it. Bill Walker and I often used to be up in the silo together when we were in high school.

Emily: I hardly remember that.

Jim: I know it. I hadn't either, until I saw that picture of the man.

Emily: I guess I wasn't really out in the barn when that was going on . . .

Jim: No, you wouldn't see that much. That's a flight through time all right.

Emily: I was probably helping with the pies.

Arthur: We'd always had a rope on the pipe and you'd make it nice and level around. Keep it level. Used to put two or three in the silo sometimes didn't they? I don't know whether it did any good or not but . . .

Willie: I don't think it did that much good.

Arthur: But you could put three in there. You keep that all tramped down and the guy that would hold that pipe—other people puttin' her right around.

Willie: Some of 'em used to run water in the corn . . .

Arthur: The gases didn't come 'til after it fermented, after it'd been in there awhile.

Jim C: The neighbors were filling silo; there was a couple of guys in the silo and they got to foolin' around and they got bettin' which one could lay down and let the corn pile on 'em and still get up. By Jesus, one guy, . . . if the other guy hadn't a been there it'd buried him up. It was comin' in faster than he realized it. He couldn't get up. The other guy had to pull him out.

Willie: But after awhile you got an awful lot of gases. It don't take long for it to start fermentin'.

Mason: I know it. You get a shot of ragweed every once in awhile. . . .
Kinda' makes you smell.

Willie: Makes the old nose run and your eyes water.

Mason: This guy looks like he's too old to be out in the field. That's probably what happened. He's too old to be pitchin' corn or drivin' teams. So they put him in the silo.

Accordin' to that rope he's got wrapped around his hand he's pretty well up in the silo too. He's up quite a ways.

The Field Chopper

The entire way of working, spending time together, and organizing a farm was revolutionized by a single machine. Carmen explains:

Fig. 77 Tractor drawing Papec silage field chopper, A. E. Schudder farm, Painted Post, New York, September 1945. Photograph by Charlotte Brooks.

We always changed hands back and forth, all the ones in the area. At least ten to fifteen farms changed back and forth. You needed about fifteen to eighteen men to fill a silo. You used to have about eight men out in the field pitching on corn, you always had a couple in the silo and you had about four or five wagons drawing—you needed quite a lot of help! It was the same for the grains—the same amount of help.

Doug: What put an end to that?

Carmen: The field chopper. When some of the people got to filling the silos with the field choppers then there wasn't enough help to do it the old way. See, when I started filling silos with a field chopper, I didn't go out and help anybody on the filling. They didn't require any outside help, or only a little, and so after a few more got their choppers everybody else went the same route. I think it was about nineteen probably fifty-three, -four, somewhere in that time Garnet Beckstedt bought a chopper. I was one of the first

Fig. 78 Mechanical corn-harvesting equipment being used on a farm near New Lebanon, New York, August 1949. Photograph by Charles Rotkin.

farms he did, I guess. And that put an end to changing works.

Doug: Was it so much faster to chop it in the field?

Carmen: Oh yeah. He could do all mine in a little over a day, at the most. He cut pretty near ten acres a day with probably three men. The other way it would have taken probably three days, or at least two days, with the whole gang—seventeen, eighteen men. They did everything and they got paid by the hour. I don't remember what they got paid!

Doug: How many years did it take for the old system to die?

Carmen: Within three years everyone was hiring someone to fill. Then I got my own chopper, where Victor Cryderman lives—his father bought one and went around filling, and so it got so that everyone started getting their own outfits. Of course, today pretty much everybody's got their own. But, at that time, there were a lot of people who went around filling silos. Most everyone had their own ten years ago, and the rest have bought theirs since.

The field chopper necessitated bigger tractors, and, eventually, self-unloading wagons, all of which were paid for with a bigger milk check produced by the herds the greater quantity of corn would feed:

Doug: Did you have to buy a different tractor when you bought the chopper?

Carmen: Oh yeah. When I got the first one I had the old M—that was big enough to pull it; the other tractor I had wasn't. I still have it!

Doug: Did you increase your herd size then?

Carmen: Yes, we got up to thirty-five or forty.

Doug: Did you increase the herd size because of the field chopper? Because you had more expenses in terms of equipment?

Carmen: We just kept getting more feed, adding on to the barn, and putting more cows in it. When I started out at first I had only ten cows. But in July we bought ten more. I can remember saying, if we get that side full, that will be it, that's all we'll ever milk. That was twenty-five.

But buying the field chopper, which more than any other machine transformed farming, almost seemed for Carmen like an afterthought:

> *Carmen:* Beckstedts got the first one and the next year I bought one because it made me mad that he didn't come and fill mine when he was supposed to! I said to hell with that! I should have got after him but I was stubborn enough not to! He drove right by my place and went over to fill where Tim Shoan is, at Ernest Rutherford's. . . . Then he came back and done mine. Ernest was chasing him, I guess, to get his filled. So then I got the chopper and I don't remember exactly what year I bought the combine.

Farmers who bought choppers hired themselves out to neighbors to do work they had all done for free in the past. If this system worked well, farmers who did not buy the machines could find and afford farmers to do their work when they needed it done. Farmers who bought the new machines could find customers, and they could pay for their machines with the fees before they wore out. In perfect balance, it worked well. But it was an unstable system, and most farmers eventually either bought the equipment or left farming. Buck explains:

> Well again somebody bought a field chopper then he worked out with it. Just like this round baler I got now. I can get work with the round baler because everybody don't have one, or maybe the farmer don't really bale that much, but he wants some baled so he will hire somebody. But this chopper business, to get back to that again, after a few years, like fifteen years, these ones that went around chopping corn—they got older, and they got just to where they didn't want to do it anymore. So it got harder to hire somebody to chop your corn, so then they had to go and buy one.

Figure 77 shows an intermediate stage; a worker was needed to direct the field-chopped corn to a wagon. These machines were eventually replaced by self-propelled choppers, for those farmers with a few hundred thousand dollars to spend, which blow corn directly into wagons. Thus, we see in this simple and seemingly inevitable evolution of machines the redefinition of social relations on the farm and between farmers.

Virgil explained:

We'd go around buzzing up each other's woodpiles. Everybody had a woodpile and we'd take the buzz saw around—it was on the back of the tractor. The farmer would have his wood in sled lengths. He'd draw it up in sleds, put it up on a pile . . . set up the saw right in the middle of it. That was a job. Then put your logs on the cradle. Some of them were that big [*circles arms*]—you'd have to roll them over. The saw wouldn't be big enough to cut them all the way through. Then you'd split it by hand. You'd cut it for the year ahead. As soon as the crops were in, that was the next job—saw the wood pile up! Get it ready for winter! The neighbors would work together on that, too. You had your neighbors in the house a lot.

The Meaning of Changing Works

I have described changing works as the collective harvest of small grains and corn. This was, in fact, the most organized and extensive example of labor reciprocity. But changing works was part of the general culture of farming, common to many specific agricultures.

Arthur: You want to talk about "changing works"? . . . I owned a nose-piece; I owned the rope. And I knew how to cut off a cow's horns. A guy bought the saw; I cut his horns off for thirty-five years for nothing. This is the way it went.

Doug: I don't get it.

Arthur: Well, I cut his horns off for thirty-five years for nothing.

Doug: Why?

Arthur: Well, he owned the saw. . . . Ten cows a year, go and saw 'em off. Thirty-five years.

Doug: You used the saw for everybody else?

Arthur: Oh, I did a lot.

Doug: In other words, you'd use his saw to work for other people and for using his saw you'd cut his horns and . . .

Arthur: Good. . . . Big deal.

Now he came in here one day about noon, right after Ruby and I were married, and he had some loose hay to get in the barn. And he said, "I just broke my rope. I'd like to get 'er spliced." Well, I ate my dinner and I went up and spliced his rope. Ruby says, "You're not gonna splice that rope with hay out in the field." I went up and spliced his rope and come back home and got my hay in, what I could anyway. And he got his hay in. Next day him and his hired man come and they worked all day at my place. Never paid 'em nothin'.

Fig. 79 Signs of autumn: cutting wood on buzz saw, November 1945. Photograph by Charlotte Brooks.

That's the way they were. We changed works a lot of times. He wanted to have my hay get in too. They cared about yours.

Willie: Because you left your hay set to go over to splice his rope.

Arthur: They don't care now. Heh?

Willie: You left your hay set so you could go splice his rope . . .

Arthur: Oh, yeah. He'd better come and help me too, hadn't he?

You always got enough back, generally. Didn't ya?

Willie: Usually.

Arthur: If you needed a team of horses, they'd let you have a team of horses. If you needed an extra horse . . . wanted to use a three-horse team . . . they'd let you have the best one they had. That was the worst job for the one that walked in the corn. Wasn't it? It took a good horse to walk in the corn all day. The next day if you wanted a team you hitched the two, the one that worked all day and the one that didn't, so you'd have a long team. One ahead of the other one. One's too tired to work. Is that right?

Willie: Yeah.

Arthur: Things have changed.

Jim Carr described how neighbors spread manure together:

> We used to have a big manure pile; he'd have one too. We'd spread with three manure spreaders and the one tractor loader. We had two manure spreaders of our own; we had one; and Frank Short, he had one; and Austin Martin had one. And when they spread down to Austin's, we'd take an extra tractor and use his manure spreader, and he'd load the manure. Then we came here; we'd use our tractor and his spreader and he'd load. Down to Frank's, he'd load and somebody used the spreader. And we'd spread with three outfits all the time. We'd probably put out a couple of hundred loads per place.

The Social Nature of Changing Works

The principal of changing works was that farmers informally organized themselves (meaning the arrangements are not contractual) to share labor. There were many forms of changing works in different regions of the country, depending on the duration of the work which needed to be done, the density of farms in a region, and the technology at a given stage of agricultural development. For example, Jane Adams describes how farmers changed works in an agricultural region of southern Illinois during the 1920s. In this region farmers grew several crops, including large fields of grain. Each July a group of farmers in a neighborhood collectively rented a thresher, which they moved from one farm to another. Given the size of grain fields in that region, it took a thresher crew up to two weeks to do the crop on a sin-

Fig. 80 Farm hands at dinner after threshing wheat on the Shaughnessy farm, Warsaw, New York, August 1945. Photograph by Sol Libsohn.

gle farm. While harvesting the grain, the crew stayed on the farm. Thus, the farm women fed up to twenty people a day for ten to fourteen days in a row, on an improvised dining room table temporarily erected in the living room. These changing works meals were of sufficient importance that, addition to the farm workers, "the dinners drew in townspeople, such as the banker and merchants, with whom the farm family did business, creating an arena in which business relationships could be amplified and given a personal dimension."[25]

In northern New York, labor sharing involved about seven farm families (the grain and corn harvests could be divided into about seven parts), who could share the technology needed to make changing works operate. The threshers and choppers were affordable by a sufficient number of farmers so that the harvests could take place during the short meteorological window. Like all examples, the North

Country version of changing works was the result of a formula including crops, weather, technology, and available labor.

Finally, there was a ritual dimension to changing works. As workers circulated from one farm to another, they ate together in each other's houses, reenacting an event in which social bonds were reaffirmed.

The tradition of changing works in the North Country drew on these elements: homogeneity, isolation, and a tradition of spending social time by eating together. Changing works brought neighbors rather than kin together, and the changing works jobs often went on for weeks. It took farmers to each other's homes, where they normally would not be invited. One farmer I interviewed mentioned his fascination, as a teenage boy on changing works crews, with the religious icons of his Catholic neighbors (this was a predominantly Scottish Presbyterian neighborhood). In these ways, the labor exchange tied people together; it created a group which addressed people's common needs for both work and culture.

Eating played a big part in the rituals. Food is a powerful basis of celebration, for it addresses one of the basic human instincts. To eat good food together is a typical form of social life, probably in nearly all cultures and all historical eras. Not surprisingly, farmers spoke fondly of these events, especially when they reflected on Sol Libsohn's photos of a changing works dinner.

Jim Fisher: This is a great scene isn't it? (See fig. 80.) The threshing crew sitting at their table eating.

Doug: Are these all farmers, or are some of these hired men? Can you tell by looking at it?

Jim: Well, yes—there is a great variation in age. I would say all of the older people are probably the farmers; these young people are probably hired men or farmers' sons.

Doug: What have they just finished eating?

Emily: Pie and coffee.

Jim: They probably had roast beef . . .

Emily: Looks like cottage cheese . . . they made their own.

Doug: Looks like they are having cigarettes too.

Jim: Oh yes.

Doug: They are quite serious. What were they talking about, do you suppose?

Emily: Well, it wouldn't be about the work they were doing—it must have been about politics. He doesn't care that he is taking the picture. Probably thinks what are things coming to anyway—making a sociology study of us!

Buck recalls:

> I can remember back years ago when they would be at the dinner table they would be talking about something that happened maybe years before that—"can you remember this or can you remember that." I remember one time they was cutting wood here and somebody said: "Come along, the bullheads are coming through the ice at Mud Lake. If you chop a hole the bullheads are coming right up through the ice," and somebody said, "Well, if we didn't have to cut my father's wood we would be able to go up there today and get ourselves a lot of bullhead." "Well," my father said, "Don't let that stop you!" So they all quit cutting wood and they took ensilage forks and they went up and they shoveled the bullheads. You cut along the ice; of course the ice was thick, and there wasn't a lot of water when you cut a hole in the ice, and the bullhead would come right up out of the hole. They shoveled them right up with ensilage forks—and bagged them. They might be talking about something like that. I am sure that they just didn't talk about work.

I asked Tom and Marie if the farmers in the photo would be talking about the day's work, or larger issues, maybe politics:

> *Tom:* Probably more politics than whose field is next.
>
> *Marie:* Really? I don't know, I always thought, when I sat around listening to them, that they were talking about what crop they thought was coming in, and what they thought when they'd be getting to the next neighbor . . .

Virgil simply remembers:

> We ate together, too. Oh my, I guess so! That was a big party then! The farmer never gets a day off—the cows have to be milked every day and this would be different.

Changing Women's Work

Some of the photographs reveal previously hidden actors: the women who prepared the farmers' feast. Changing works was essentially a male domain, and although the intention was not to dominate women (rather, one might say, to carry on the normal order of things), the shared labor brought men together, nourished their social selves, and consolidated their social power. The system through which men extend their own power at the cost of shared power among other social actors has been described as "patriarchy."[26]

North Country women experienced its effects, relative to changing works, in some variation or combination of three ways. For some women, preparing the changing works feasts involved social enrichment. Women overcame the loneliness of the farmstead preparing food with other women; and for some, hosting workers

Fig. 81 Farm women dressing chickens, Brooklea Farm, Kanona, New York, September 1945. Photograph by Charlotte Brooks.

enhanced their reputations and their sense of self-worth. Some women resisted the role of cook and host or simply did not perform very well, and, finally, a small number of women participated in the work crews as equals with men, even while maintaining their traditional female roles.

SOCIAL ENRICHMENT

When the subject of changing works came up in discussions with farm women, most spoke of the social enrichment they experienced during this period of increased work. There were two reasons: women often worked with other women (relatives, neighbors, or special friends),[27] gaining social solidarity similar to what men gained in their participation in work crews. Second, successfully organizing and accomplishing the meals brought respect and status to many women.

Cooking was a routine part of women's responsibilities, which became much larger when the changing works crews visited their farms. They rose before daylight to bake bread and pies on woodstoves and worked through most of the day to cook fresh or canned meats or newly harvested vegetables for ten to twenty hungry workers:

> *Mary:* We made potatoes and meat—salt pork was one of the things they always had. And, believe it or not, I always liked it. Sometimes you'd get chicken. You'd get vegetables—of course we always had our own vegetables. You just cooked a regular farm meal.
>
> We saved ahead for this. We had roast meat; we had chicken. We used to put up pork. You can't keep pork very long. But there was a kind of a pickle we used. It was made with brown sugar and salt. We put the fresh pork down in it and it would keep it a real long time. Deserts were pie, cakes, all different things. They were great cooks for desserts. Puddings. The men liked pies. We used the long blackberries; we always had lots of long blackberries. Men always like apple pie. Virgil's father—he'd always have apple pie. There would be all kinds and he'd ask for apple.
>
> The garden was not in yet; you saved pickles, relishes, canned goods. You made fruit pie—you could not get by with cake!

Tom Lee had recently talked to his grandfather about his memories of changing works. His grandfather had remembered:

> The women would put on a meal and each woman didn't want to be outdone by the others. We got fantastic meals!

While in northern New York, women did not generally circulate with the men from farm to farm, there were many informal arrangements in which the labor was shared. Marie remembered:

> When the farmers came for threshing the wives got together and helped one another with feeding the men. Not, all of 'em, but there would always be two or three that would cooperate together.

Mary remembered it similarly:

> The women would come and help set up, set the table, clean up. I had special friends—don't we all? . . . You get to know them, know their habits. I grew up with a lot of their families.

Jim and Emily recalled:

Fig. 82 Farm hands at dinner after threshing wheat on the Shaughnessy farm, Warsaw, New York, August 1945. Photograph by Sol Libsohn.

Emily: The wives didn't change, but generally there was someone else who helped.

Jim: Often, there were mother-in-laws. Right in our house, my mother was here, and usually on the farms that were large, and had a lot of changing work and big meals, there were neighbor women who helped. Hazel used to come help Mother.

Emily: I think that's true and even in our generation, we would like to help each other. Ruth Walker would come over, or I would go help her.

Jim: Of course, we had two. Mother was here and you. So it depends. A lot of them were mother and daughter-in-law arrangements. Usually they were double houses.

Emily: But we didn't change works—that made too many people in the kitchen!

Jim: When you read Emily's diary, we were feeding twelve to fourteen people all the time anyways, with the kids and their friends. I can't believe the people we were always cooking up for!

Emily: After mother was gone and I was alone—I remember when we would have changing works and we were expecting a silo crew. Murray was in first grade—he had gone to school and Diana was just lost; she didn't have anybody to play with; the other little ones were too small I don't think even Stephen was here then. But, anyway, Aunt Bess came up to help and Ronnie Day was just as lost as Diana, because he didn't have anyone to play with either, so he used to come up and play with Diana. We would bake the pies—Aunt Bess was a great pie baker and we would put them out on our screened-in porch to cool—they were on a table which was very substantial until somebody leaned on the end of it . . . so the children had been playing on the other side of the hedge and they got curious, and they came running up on the porch. I didn't think anything about it, but they wanted to see what was on the table, and they took a hold of it and leaned on it. We heard a noise and went running out, and they were gone but the pies were upside down on the floor. Aunt Bess was such a practical good little person and she said "flip those pies right back in those shells—we will never mention it." But I wouldn't have dared; I don't know how she did it. I would have never been able to flip a pie back into its plate. Of course the pies were sort of still in the pie plates—they hadn't completely fallen out!

Jim: Why I think Aunt Bess baked a pie for Uncle George every day of their married life—practically—and he lived to be eighty-three!

It is difficult to judge the attitudes toward gender-based work roles which were in full bloom fifty to a hundred years ago in the context of modern consciousness. In general, most of the women and men I interviewed remembered changing works in terms of the important tasks that were completed through the interdependent, though unequal, efforts of men and women. There was no general sense that the gendered division of labor could have been organized differently.

RESISTANCE AND INCOMPETENCE

Not all women accepted roles that put them in the kitchen in service to the work crews. Marian's comments in the following exchange are instructive:

Doug: Were you involved in making these dinners, Marian?
Marian: Well, yes, I always had to help.
Carmen: They had a lot of fun with it.
Marian: Oh, it was a drag!
Doug: Perhaps I am idealizing it?
Carmen: The crew was usually there for dinner and supper.

Marian: Oh yes . . .

Carmen: Oh, the women got a kick out of it!

Marian: Oh, baloney! I really created a stir in the community when I said I wasn't cooking for the men any longer for silo filling. I really got a lot of flack for that! Everybody brings their own lunch now, and nobody seems upset about it this year.

Carmen: A lot of the women won't even cook for their hired man anymore!

Marian: Very few . . .

Carmen: In fact some of them don't even want to cook for their husbands!

Marian: I don't know of anyone who packs five lunches for their men, either! Six days a week. People don't believe I do it.

Carmen: If you have to make one, you might as well make five!

This exchange is interesting for several reasons. First, Marian, who spent much of her married life doing production jobs on the farm, rejected her role as cook and host, even when doing so bucked community expectations. Second, Carmen, who believed he understood Marian's role in their farm enterprise, completely misunderstood her sentiments regarding this matter.

One can assume that Marian's attitudes and behavior were experienced by many women. Jellison, for example, cites a 1913 study of farm wives, which indicates that "a frequent complaint among respondents [farm women] had been the extra work that women were required to perform for non–family members, particularly for hired help."[28] In that the survey Jellison cited took place during an era in which changing works was the common means through which the harvest was accomplished, the data suggest we ought not to take women's attitudes toward changing works for granted.

In another example, Mary, one of the most traditional women I interviewed, spoke sharply to her husband about a changing works incident which happened several decades before:

Mary: I remember one time they got through threshing at Virgil's father's. His mother and I had dinner all ready. Well, they were going to move in on this other lady. "Why?" I said, "she doesn't even know they're coming!" So we got our dinner all ready, took it down to her place, put it on her table and they ate down there before they ever went out on the fields. They could have just as well eaten here! But, oh no—they had to go down to get set up.

Virgil: We wanted to go down there and get set up before noon. See, we had got done with dad's threshing the day before, but we didn't get done before dark, and so we had to finish up in the morning. So we wanted to get down there and set up before dinner.

Mary: Oh, I don't know what the old woman would have done! She'd have had to start right from the beginning to cook for that amount of men . . . there would probably be ten altogether, maybe twelve.

This exchange shows that the male farmers never questioned the value of their work in the context of their wives' expected contribution. It was taken for granted, and their inconvenience was not of great importance.

Not all women prepared the sumptuous feasts the farm women describe above. Some women had fewer resources or were less well organized, or simply were less competent cooks.[29] Others were maybe too greedy to give away their best food to the work crews harvesting their crops. Or some may have cooked with less enthusiasm because they did not accept the expected roles. Several farmers discussed these ideas:

> *Doug:* Was everybody a good cook?
>
> *Mary:* I never heard it if we weren't! I didn't see anybody turn anything down! There were mostly good cooks up here at that time.
>
> *Virgil:* [But] there were some places, you had to have an excuse to go home for dinner! I'd have to go home to feed the horses or something!
>
> *Mary:* He'd have to go home and feed his hens or do something! It was just plain unbearable!

Mason and Arthur had the following to say:

> *Mason:* There was an understanding that those with money put out less food. The better off didn't put their money on the table. The poorer farmers saved their best roast and brought out their best potatoes.
>
> *Arthur:* Harry Pierce owned the biggest farm and he was the cheapest S.O.B. in the neighborhood. I've seen him kill a chicken at eleven o'clock and have it on the table by noon!

Like the variation in cooking, the conditions in the farmhouses themselves ranged from organized and clean to disorganized and dirty:

> *Mary:* That's what would turn you off! The conditions of some of the farmhouses. But I don't think that people who live that way care. I honestly don't. If they cared they would do things differently. So I always figured it wasn't bothering them; they lived in it all the time, so I'd have to swallow my pride and go right along with them. It wasn't always easy!

It is, of course, impossible to know what would be behind these variations in farm quality. It might be tempting to call it a form of resistance to the expected role. Or it may simply be a matter of some farmers, both male and female, having less interest in, and competence with, domestic affairs. In either case, the effect was the same: the domestic side of changing works barely performed to minimal expectations.

SWITCHING ROLES

Some women worked on the crews alongside men, and on rare circumstances a male took a job in the kitchen. These gender switchings, were, however, uncommon.

Marie [looking at the photos of threshing, figs. 53–61]: I used to do that too. I used to work for the neighbors. When they got ready to thresh I would stook the grain. I used to throw it up on the thresher (figs. 57, 60). I know I remember that.

Doug: Did you also haul the bags of grain into the barn?

Marie: No. I usually worked out in the field. I threw it on the thresher.

Marie [studying fig. 80, showing the men around the table]: I used to hate this. I was the only girl in the crew other than, there was a farmer who had a daughter who used to drive tractor. But I was always pitching with the men and stuff. And come dinnertime, I didn't like it because I'd be the only female there. And it bothered me.

Doug: I never heard of another woman working on these crews.

Marie: Well, I don't think there were any other girls down around there who worked on the crews. No, I was the only one.

Tom: I would say it wasn't very traditional to see it. The neighbors she worked for were there last people who threshed that way.

Marie: The last ones that went out; the last ones who had a threshing machine.

A handful of women did both men's and women's work. Buck Wagner's mother, Gladys, was one of these women. She was widowed in early adulthood, took over the family farm, and ran it successfully for decades. In the neighborhood this gained her respect and even a degree of local fame.

Buck: She would cook up a whole complete turkey dinner and would have it all ready for the crew when they came in . . . pies and everything.

When my father passed away I was three or four. She always went to the barn [milked the cows] and drove a tractor. She could drive tractor all forenoon and she would come to the house maybe ten minutes before the rest of us would and there would be a hot meal with meat and potatoes . . . gravy . . .

Carmen [Buck's son]: I don't know how she did it!

Buck: She must have got it ready sometime or other before . . . I don't know just how she done it, you know. In the morning it was nothing to see two or three cars out here for breakfast! People would stop in just like you are here tonight. If she was here you wouldn't get out without having a good-size lunch. She was hurt if they didn't eat. We had the first television in the neighborhood, around 1950 or '51. On Friday nights the neighbors came in to watch the fights. We always had a party.

I asked Willie about these matters:

Doug: What woman really wanted to work with the men out on the crews?

Willie: Their place was in the house—getting the meals.

Doug: What if they were a rebel?

Willie: They could start getting a meal ready, taking three or four hours to get that meal ready to serve. You see they are just cleaning the chickens in that photograph (fig. 81)—maybe that is in the morning when they are doing this. The other guys are out milking—see the windows are dark. That is probably their dinner.

Doug: So in other words—they need to be in the house working.

Willie: Right.

Doug: If they were out in the field nobody would get their food?

Willie: Right.

Doug: But Buck Wagner was telling me his mother would go out and work in the fields.

Willie: She loved riding tractor!

Doug: I heard that. She liked to be out working.

Willie: Oh yes, they didn't come much better than her.

Doug: So there was some exceptions.

Willie: Yes there were some exceptions. She could get the meals *and* do the work. While they was cleaning up and getting things ready to go for the next day—why she would have the meal ready for them when they got into the house. She would have had to have gotten everything ready—all she had to do was heat it and put it on the table. She was a great person.

But while rare women took on the expectations of both male and female roles, and were respected for it, the even more rare instance of a man called upon to help women brought him ridicule:

Buck: See my aunt was a poor cook—that is the one that lives in Florida. They lived down the road here and Lorni Cryderman used to live up here at the Cryderman place. He was probably seventeen or eighteen at the time. Well, when they filled the silo at Ralph Dunn's place—Ralph would always say to Lorni at about 10:30, "You go in the house and help Eileen!" So she would give him a pan of potatoes, and I can see Lorni yet sitting out there on the steps peeling the potatoes and everybody going to the field with the wagons hollering at him, "How are you doing, Lorni—are you going to have the stuff ready by dinnertime?" He would be just seething, you know. A lot of the women used to go down and help Eileen because she couldn't manage it alone, whereas like my mother, and my father's mother were here, and there wouldn't be anyone else that I can remember coming to help.

To draw this discussion to a close, the shared labor known as changing works played out in the context of traditional roles. Undoubtedly, changing works strengthened men's roles and their solidarity across farms. In that way it was part of the patriarchal character of farming. Many women I spoke to accepted their contribution to changing works because they saw gender-based roles as natural, and they understood that only through interdependence would the work be done. Some women overcame the isolation of the farm when they helped others cook, and some women used the work to see friends whom they were not often able to visit. Other women saw that when they successfully organized and provided for the changing works feasts, their reputations grew in the neighborhood. A small number of women, however, resisted their expected roles in the changing works harvests. In general, however, the organization of the labor exchange was based on gender.

On the Work Crews

I asked farmers about their relationships with other workers, and whether they got along equally well. Several remembered jokes played fifty years before:

> *Jim Fisher:* We were threshing and George told me to put a rock in Carmen's sack. See, we had to carry the grain sacks to the barn. So we put the rock in the sack, and, oh, it was heavier. When Carmen dumped the load he found the rock, and we heard about that when he came back!

Arthur and Willie described how crews treated workers who didn't work up to the expectations of the group:

> *Willie:* Well, they were dealt with what you'd call "rural humor." I remember once we stopped for lunch, and Bill filled up his tractor with gasoline—we were using the tractor to run the ensilage cutter to fill the silo. Well, somebody drained all but a little of the gas out, so when we started up after lunch it run just a little while, and then the chute got plugged. Oh, he had a hell of a time unplugging it. And he knew somebody had done it just to get him!
>
> *Doug:* Was he not as hard a worker?
>
> *Willie:* He was a bachelor farmer and he lived with his mother. It was more that people did not like the way he talked about his mother. Oh, he got it a lot of times. I remember, one time somebody put ripe tomatoes in his boots. We took our boots off during lunch. They put the tomatoes right down in his toes and damn he was mad! Then another time somebody took a corn cob that had blight and ran it across his face. He chased that guy all the way to the river.
>
> See, if it was real damp, there would be these big corn boars—corn worms that would rot the corn. It would actually turn black. If the corn is not completely dried, the worms will grow on corn and that makes them slimier than hell. You take couple of those—some of them get to be a good size—and drop

Fig. 83 Threshing kidney beans, Kanona, New York, November 1945. Photograph by Charlotte Brooks.

them in someone's boot. They stick their foot down in there after lunch. But it was all in fun. Nobody ever got hurt. Now one of the other neighbors threw a skunk into the silo when a man was working in there. That wasn't funny. It was a hell of a mess!

I asked Jim Fisher whether peoples' idiosyncrasies ever stuck out:

Jim: There were a few particular ones that, you know, were always chuckled about. Everything would have to be so-and-so for his team. Then there were always the more easygoing ones. But that's just part of the fun, of course. The farmer who lived in your house—Lester Joyce—he had to have his pie served before his dinner!

Buck remembered the social side of the work crews:

We used to like the beer pretty well, so we would stay at the threshing machine because the thresher is in the field.

If it was hot, most farmers, when they were working, if it was hot they would have a beer. They didn't drink to excess but they appreciated a bottle of beer once in awhile. The only thing is, the ones out in the field didn't get very much. If you stayed at the threshing machine you could get all the beer you wanted—so that is where we stayed most of the time. Then my neighbor up here she had those pretty flowers—red, blue, they grow up in a cluster—peonies maybe. I was drawing the oats and we to back up into the barn and unload the oats into the granary. We had a Ford truck—I can't remember what year that was now, but it was a new truck. Along about five o'clock at night we had a little bit too much to drink. I was pulling the truck up into the yard and I wasn't watching where I was going—I was running the bumper and the radiator right over the top of her peonies and the fan was clipping the peonies right off of her plants.

Doug: She probably wasn't too happy about that!

Buck: We had a good time.

Doug: Now I wouldn't have guessed that there would be a bottle of beer out in the field like that. You see these photos and they look like they are straight laced and serious . . .

Buck: Most of them didn't have that much—they would have two bottles all afternoon. You took water out to them you know. You wouldn't get sleepy working the threshing machine because you had to keep up with it. Oats were coming out of the machine and you had to keep up with the machine. . . . If you didn't keep up with the machine you went down the road.

The End of Changing Works

The labor exchange serviced state-of-the-art machines. When machines were developed that did not require large labor crews, the system, although more than a century old, quickly fell apart. The old system did not mix with the new. All farmers had to participate for there to be a sufficient number of workers to service the existing machines and acreage. But beyond that, the change represented a cultural shift. Suddenly changing works became part of the past, a form of life to be put to pasture with the obsolete horses.

Farmers who did not buy choppers or combines hired neighbors who had previously worked by their sides. The farmers who purchased the new machines, in turn, paid for them by hiring themselves out to their neighbors. This increased work for entrepreneurial farmers, and the extra work wore out their machines, perhaps before the final payments were made.

Jim Fisher, a farmer who made the transition to new technologies more slowly than many of his neighbors, described the other end of the arrangement:

Doug: Carmen talked about getting the second field chopper in the area, and he said when he got his machinery he had to work out to pay for it. So, if you didn't want to buy one, you might hire Carmen to do your silos.

Jim: Yes, we did that, avoiding the big investment. Then it got more difficult to find someone who wanted to do that. Because, in the meantime, usually they got bigger themselves, and where they had been used to doing their neighbors' along with their own, they increased to the point where they were so tired with their own routine that they just couldn't think of assuming that added work and expense, even though they were getting paid for it. It was wear and tear on their equipment, and it was wearing out faster. So it began to get more difficult to hire someone to do it. Carmen still continued to change works with some of his neighbors, like Timmy Shoen and Bill Fisher and Bill Day, but eventually Bill Fisher and Bill Day got their own equipment. Carmen filled our silo for quite a few years. Then it got so that when you called they'd say, "Well no, they just had gotten so much they didn't want to do any more corn." Max went that route; he combined our corn one year and I thought, "Oh, this is great, I'll plant more corn next year." But when it came to be halfway through the season, he said, "I'm just not going to work out this year, I've got too much to do."

In these various ways, the practice of trading labor partially mixed with new entrepreneurial activities. Within a few years, as Jim Fisher recounts, it became nearly impossible to hire neighbors with the new machines. The farmer was left with two choices: either modernize or leave farming. Since the new harvesting machines required new auxiliary machines (larger tractors, self-unloading wagons, new silos, for example), it was necessary to increase one's income to remain a farmer. This was done by increasing herd size, increasing milk production with the existing herd, and growing more crops on more acreage, either purchased or rented from neighbors who did not modernize. Many, of course, did not make the change.

Farmers' Attitudes to the End of Changing Works

It is always difficult to assess the past, especially when changing social relations are mixed into histories of the machines and tools. There is no doubt that modernization represented, in the farmer's mind, easier work, even if it redefined what it meant to be a farmer and a neighbor:

Doug: Did the farmer welcome the change?

Virgil: Oh yes, I guess so! All that work is no longer necessary!

Mary: I missed people coming! The farmer never gets a day off—the cows have to be milked every day and this would be *different.*

Doug: When changing works stopped, were farmers glad that there was a lot less work to do, or did people miss the old neighborhooding together?

Buck: Well I am sure that there were people that missed the dinners. I mean back in those days the dinner was the highlight of the day. You done the work and didn't think nothing of it—you changed works and didn't think nothing of that, but the highlight of the day was the dinner—the pies—because you got a dinner like you would get at Thanksgiving or Christmas. The women back in them days really knew how to cook more than a woman does today because they did it all the time.

Carmen, who had been a leader in the technological revolution, said:

One thing about the system that replaced it, you hardly ever see your own neighbor anymore. You only go up and down the road—you all do your own work, and you don't very often see them, as far as neighbors. You know, really—how often have we seen you this summer? Only when we go up and down the road—really.

It's an awful thing to say—but you could live over at that place of yours for twenty years and nobody would even know who you were. It shouldn't be that way. Just like Bakers over here—their land joins on ours. We wave when we go by and they wave when they go by—and that's about as far as we see them. We're lucky if we talk to them once or twice a year. And it's too bad. The same with Rutherfords. Barbara Rutherford up here—I haven't seen her in two or three years, and they're only a mile from us here. I don't know when I've been in James's house—it must have been years since I've been in my neighbor's house over here. I might go to the barn for something and they're over here for parts lots of times; it's things like that, you know; its not what it used to be, that way.

Doug: Has that been a gradual kind of change, or was there one point when everybody got their own equipment and stopped working together?

Carmen: I think that was probably a part of it, really. There was no need, no call, really, to go see them. . . . I don't think anyone has anything against anyone—you just don't have any need to be there.

I asked Carmen whether it would be accurate to say that farming had gone from more of a group experience to more of an individualized experience.

Carmen: That's about what it amounts to. A lot of hours alone on the tractor.

I think that maybe years ago, or even now in Canada if somebody moved in—in the winter they had card parties and that neighbor, whoever he was, would be invited to it. And they'd turn around and have a card party and invite everybody back—that's just the way it went. Hell, Mother and Dad used to go to card parties twice a week, all winter long. They'd drive with the horses cause they couldn't go any other way—the roads weren't plowed at the time. There'd be probably fifteen or twenty couples. They'd play five hundred, or whatever it was—mostly five hundred—they'd have coffee and little desserts—and they'd have that twice a week! That went on until two o'clock in the morning. Cards, then the lunch visit, and they'd have to drive home after that. Of course, they weren't getting up in the morning, weren't milking so many cows—

Marian: There wasn't any pressure!

Carmen: If you go back to the changing hands, then you buy a chopper, the next guy buys a chopper. Maybe that changed it more than anything else in terms of neighborliness, really, when you stop to think about it.

Doug: As the farms became more unequal in size, did farming get more competitive?

Carmen: Maybe so. I remember one time after I bought the old M tractor I was going past Art Thompson's down there. They were threshing. I had the baler. I don't know if I was baling straw or what. They didn't tell me, but I found out about it afterwards—they said they couldn't see why that Canuck had to have a big tractor like that! Because they all had little Fords, H's, and stuff like that. And one fellow told me afterwards, they were standing there threshing: "Why in hell a Canuck would buy a great big tractor like that with a small farm?" They used to think that was a big tractor, an M tractor! And now that's just the tractor we use for pulling wagons.

It's all changed. When I was growing up we had barn raisings. I've been to them. I remember my grandfather there put up a new barn; they got the lumber cut in the wintertime for the framework. Then they got them all marked and laid out together and pinned them together, just like the old barns used to be.

We would all pitch in and go and help. Everyone wasn't so busy then—it didn't seem like it, you know. Oh, they had time, or something, I don't know what it was. Now they got so damn much going on, you're going so far and so fast and furious that everybody is going with their shirt tail straight out and don't have time to say hello, hardly.

I asked Arthur if neighbor relationships began to change when farmers stopped working together.

Arthur: You don't have any neighbors. You don't have any neighbors now.

Doug: You have people who live close to you?

Arthur: They ain't neighbors of mine.

Doug: What does it mean to be a neighbor?

Arthur: Well, they used to play cards with us at night. Come over. They'd always come see ya everyday or you went there. You don't see that anymore.

Once in awhile you see somebody through a hard time. You know?

Willie: You gotta go see which one had the best cider and that . . .

Arthur: Oh, yeah. Which one had the best cider. And the cards.

Doug: So when you were working together you were socializing together.

Willie: Oh, yeah.

Doug: Well why didn't you keep socializing after the work got easier?

Arthur: There wasn't so many people working . . .

Willie: Everyone had to go to town and spend the money they was making.

Arthur: You know what came—television came, so everybody stays home, you know, and they don't go anywhere. Well, of course, that happens to be the same time that the TV comes in; the combine comes and about then the field chopper . . . that's about when you got your milkin' parlors, right? Is that right?

I ask Jim Fisher, a young farmer when the transition was taking place:

Do you remember your attitude—did you feel that the new way was better? Or did you feel that maybe this new age was heralding in something that might not be desirable?

Jim: We were all very enthusiastic. We looked at the technology as releasing everyone from a great deal of drudgery. And those early tractors were pretty simple compared to the tractors now. They were easy to maintain, and the equipment was small. They would pull a two-beam plow and it ran on Kerosene. It was a very economical tractor. There was a small tank for gasoline for starting it and then a large tank to run it with for the kerosene, a cheaper fuel. But all fuel was cheap then!

Doug: How did people feel about that change?

Jim: Well, I think they weren't too nostalgic about the change. Actually, it sounds romantic and good, but there were some bad features. By the time you got all the way around in the fall, you were usually working in the snow and the rain on a cold day, and tramping through the mud, and you'd be numb practically when the day was over. So you remembered those things and when you could think, "Oh, if I had my own machines, and could work when the weather was nice, and get my crops all in in a week, wouldn't it be great! And I wouldn't have my month going around to all these farms." It usually took a month or so—you would do the grain in August, and you would have probably ten days of doing grain, and there would be a break, and then about the tenth of September it would begin again and it wouldn't be done until the middle of October, anyway, or later. There would be some pretty miserable days in there. So, whereas the changing works *sounds* great, and was fun a lot of the time, there was that element of wanting to do something better and more efficient, and when you saw this amazing machine come in that would

do all of your heavy work with none of the pitching by hand, it was a temptation.

And the scene had changed with the older people not being able to work. We went through a stage where I can remember Dad saying, "Well, the changing works aren't like they used to be." It used to be the farmers changed the works themselves, then it got to a stage where they all sent their hired men—and you had a different group of people. They sent their hired men as they got older and the work was hard. I can remember feeling . . . I had forgotten that . . . but I can remember in our lifetime even, feeling that the whole scene changed and the hired men didn't have the same attitudes. Some of them were Indians from Hogansburg that were employed by the farmers around, and they tended to stay a little bit by themselves. Then there would be a few farmers' sons, but more of the hired type of workers. It was still a lot of farmers. But then as they moved up to a little larger step and got, say, two

Fig. 84 Exhibition of a corn chopper, Flemington fair, Flemington, New Jersey, August–September 1951. Photograph by Todd Witlin.

hired men, which we did, then the owners tended to let the men go on the work routine.

Doug: As I listened to the tape of our first conversation, and also to Carmen and Marian talk about it, I thought, well it seemed as though farming was something that the community did together . . . and then all of a sudden you did it very much alone.

Jim: That is an observation that someone from the outside can see much better than we who are living in it. And if you talk to these farmers that are large active farms now I think they would feel that way.

Explaining Changing Works

The key to understanding why labor reciprocity worked, Mason said, was that a day's labor was not worth a great deal on the preindustrial farm. So it made sense from a cost-benefit perspective that a farmer gave several days of his own labor to gain what he could not pay for—the collective efforts of his neighbors to work with machinery he could not afford alone. From this viewpoint, it was rational to exchange.[30] To manage a dairy farm with the then-existing technology, one had to become part of a group enmeshed in complex but largely informal exchange relationships.

From a cultural perspective, we might expect that participants' social identities led them to actions which were consistent with the needs of the group.[31] Farmers were farmers, and little else. This occupation was actually a lifestyle—a master identity. For a long time this master identity included exchanging labor with one's neighbors, and in these moments the farmer played out his role, which included how others who were his peers defined him. These are not insignificant pressures!

The importance of the labor exchange was largely due to the fact that farm work was generally solitary and changing works allowed for regular social contact which the farmers wanted and enjoyed. Aspects of the exchange, such as eating and joking, became what Durkheim referred to as "secular rituals," which gave rise to "collective conscience."[32]

But for all their regularity and importance, the events and rituals were also complex, contingent, and often inconsistent, more accurately described by the analysis of exchange by John Davis.[33] All farmers had stories of bad cooks, lazy workers, or arguments over small details of the arrangements. Farmers used practical jokes partly to rebuild cultural order. By Marian's and other women's account, we see that not all women embraced or even accepted their roles in the changing works scheme.

Finally, Alvin Gouldner argued that social exchange can be explained because it is simply a fundamental aspect of the social order.[34] To find occurrences like labor exchange is not the exception, but the more general rule of social life.

Changing works is best understood from all of these perspectives. Certainly farmers calculated the worth of their contribution and saw that giving more than they were getting was good business. But the spirit of their comments points to explanations more in tune with Durkheim's understanding of the currency of ritual. Consider the farmer who said, "You looked forward to it, even though it was hard work. You caught up with your neighbors."

Some farmers gave more than they received (they participated fully even though their land holdings were minimal, and thus they received less of the group's efforts than did the others) because they wanted to be part of the group. The group's norms existed largely outside the larger culture of individualistic, competitive capitalism, even though the farms were actually competing with each other for a share of the market. As noted, the work was organized on a noncompetitive and egalitarian model. The work integrated the group, reaffirmed collective values, and created the basis for shared understanding of experience and the stories which represented it.

The ceremonial aspects of the work exchange existed in both the male and the female spheres of work. In changing works men and women expressed their separateness as well as their interdependence.

Of course, explaining changing works does not address the question of what was lost when it was finished. When formal labor exchange ended, American farmers perhaps became more American, in the sense of their heightened individualism and entrepreneurship. They quickly forgot the advantage and pleasure of working together, of building community through the nitty-gritty of shared work. In the final years I was drafting this book I interviewed family dairy farmers in Switzerland, France, and Holland. I was interested to learn that only in America had the technological change so pointedly eliminated the communal aspects of farm life.

Gendered Worlds

I have emphasized the impact of changing technology on market-based agriculture. This has led me to focus on men's experience, since men have generally been responsible for market production. This orientation has had the effect of telling mostly one side of the story.

The gender-based division of labor in dairy agriculture, however, existed in a wider context.[35] This wider context includes gender roles in other agricultural systems and the question of how productive technology—generally run by men—has a complex, often competitive relationship with domestic technology—generally run by women. Change in either the domestic or the productive technology affected the other, in often unanticipated ways.

Material and Social Conditions, and Divisions of Labor

Agricultural systems produce crops and animals based on the land fertility and arability, available water, general environmental conditions, and technologies. Production for market is influenced by transportation systems and population densities. Social factors, such as ethnicity, religion, and regional customs add their influence. The social organization of agricultural production—including the relative size of farms, influences all of the factors listed above. The influence of gender on the division of labor is part of this interworking of material and social forces. There are differing views of the "prime movers" that lead the various elements of the system to take their characteristic shape.

Fig. 85 Dinner time at Mr. Hercules Brown's home, Somerville, Maine, February 1944. Photograph by Gordon Parks.

For example, several scholars see patriarchy and sexism as the defining element in the model.[36] For example, Deborah Fink describes an agricultural system in mid-Nebraska in the late nineteenth century to the 1940s, in which the division of

Fig. 86 Spectators, Reinhart's Auction, Canajoharie, New York, August–September 1945. Photograph by Sol Libsohn.

labor defined women in an inflexibly subordinate role. Rigid and oppressive gender roles were an aspect of a punishing agricultural system. Fink writes, "imagine living in a place without beauty, without diversion, without light, with only work to fill out each day. . . . In Nebraska, women encountered a bleakness they had not known before."[37] Though the land was fertile, it was semiarid. Crops replaced the complex ecosystem of the prairie with monocultures that invited infestations of grasshoppers and other insects.[38] The farm system was precarious: only about 30 percent of the farms established in 1900 remained in business by 1910.[39] At the core of the portrait drawn by Fink, however, is patriarchy and sexism. The system-

Fig. 87 Spectators, Reinhart's Auction, Canajoharie, New York, August–September 1945. Photograph by Sol Libsohn.

atic oppression of women and children by the male farmers, and even male violence toward both spouses and children, was the system's defining element.

Fink explored the effects of patriarchy in another case study of agricultural division of labor, situated a hundred miles or so to the east and with a more contemporary historical focus.[40] The fertility, rainfall, and soil arability in Iowa led to an agriculture that evolved from diversified family farms to industrially organized farms specializing in corn, soybeans, and hogs. The region in which this case study is situated features some of the richest agricultural potential in the world.

Fink evaluates how technological intensification influenced the gender-based division of labor. Her analysis of what she calls the "dual economy" details how household production by women and children provided for living expenses, while male-dominated market-oriented cropping produced income for technological improvement, land, and housing. Fink's view, common among rural historians and sociologists, is that technological intensification on farms leads to increasing marginalization of women. Her explanation of how egg production moved from women's to men's domain is a case in point. Prior to World War II, on farms in several regions of the United States women were responsible for poultry flocks and the eggs they produced. The income from the eggs and poultry was "women's money" (often denigrated as "pin money"), bartered for the groceries needed to supplement self-provisioning and also traded and sold to pay for services such as doctoring and even, in some instances, new land and equipment. Fink and other scholars make the point that the egg and poultry system created an independent sphere for women which allowed them a degree of power and influence.[41] During World War II, as dried eggs became a staple of the U.S. armed forces, agricultural policy makers formulated policy (and manipulated public sentiment) to replace these tens of thousands of independent egg and poultry producers with factory enterprises, eventually located many states away from where poultry had been an integral part of a diversified agriculture.[42] The transition of farm to factory had the effect of moving a productive activity, and the social power it implied, from women to men. As in Fink's first case study examined here, the defining feature of the agricultural system is patriarchy.

A second perspective suggests that the gendered patterns in divisions of labor may be more fundamentally associated with the technologies, crops, and animals of a given agricultural system. Jane Adams's study of gender and division of labor is an example of this perspective. Adams's case study is situated in far southern Illinois, an area with abundant rain, fertile land, rolling hills, and a tradition of small farms. The farms were diversified, growing row crops, fruits and vegetables, and several kinds of animals. The region varied considerably in wealth and ethnic background, but the range of tasks due to the diversity of the farm enterprises was great. As a result, Adams argued that the division of labor had not marginalized women: "Unlike the women Fink interviewed in Iowa, no woman I interviewed saw her work as a 'sideline.'"[43]

Nor did Adams find the "dual economy" of pre–World War II agriculture described by Fink: "Work aimed at earning a cash income and work that provided daily needs were not sharply distinguished, just as child care was not a discrete body of activities. Similarly, work that was appropriate to men and to women was, in

Fig. 88 Canning tomatoes, September 1945. Photograph by Charlotte Brooks.

many cases, not strongly distinguished."[44] Women were expected to work in the fields as well as in the home. This portrait is developed in the memoir of Adams's informant Edith Bradley Rendleman, a woman born just before the turn of the century who spent her entire life on farms—her parents', her husband's, and, as a widow, her own.[45]

As in the areas studied by Deborah Fink, World War II provided an important turning point in Adams's farm community. Before the war, women's working roles were essentially equal in importance to those claimed by or assigned by men. After World War II, however, "The same processes that encouraged men to adopt a more rational, scientific approach to their enterprises fundamentally altered women's relationship to the farm, separating a highly capitalized agriculture from domestic production and consumption. For the first time farm and household became functionally disconnected."[46]

By the close of Adams's study in the 1980s, only two of the seven farms she had studied in depth remained in farming. The technological intensification had consolidated farming in the region to a handful of large farms and had moved much farming out of the region, impoverishing the rural culture.

Adams suggests that a farm system may differ in its social organization from another with different material and social conditions. In one system patriarchy may reign; in another its effect may be more muted. It is impossible to determine why some farm systems were more egalitarian in the ordering of work, but it would be reasonable to suggest that greater diversification in Adams's agricultural system led to a wide range of tasks that could not reasonably be organized around rigid gender stereotyping. Women were more available during some years or months than others, and most likely oriented to tasks that permitted them to remain close to the homestead, where they were responsible for child care and other domestic tasks. Beyond the pressures afforded by these responsibilities, gender assumptions influenced but did not irrevocably stamp the division of labor.

The idea of heterogeneity in farm technology and gender flexibility is supported by Nancy Grey Osterud's study of farms which occupied areas roughly the same as those represented in the SONJ images used in this project, during the end of the nineteenth and early twentieth centuries.[47] Osterud studied diaries, letters, and other documents that allowed her to understand not only how tasks were assigned, but also how men and women felt about the worlds they occupied. Osterud's farmers were established in an agricultural system in which small dairies produced butter as the main cash crop and, of course, grew the food for both their cattle and draft animals. The farms averaged eight to ten cows.

Osterud's community was characterized by what she called a "conventional division of labor based on gender" but also shared labor, particularly as concerned milking, cleaning the barn, churning and packing butter, and taking care of animals. She writes, "The allocation of tasks between men and women was flexible. It varied from one family to another, and within families over time, depending on that mix of choice and necessity we call custom. . . . families [sometimes] departed from the conventional pattern simply as a matter of personal preference or family tradition."[48]

Because of the flexibility in assigning tasks, and the typical tasks women performed in milking and butter making, "the division of labor by gender on local farms did not correspond with the difference between subsistence and market-oriented production."[49] Men were reported to help with the washing and even to sew their own clothes. Women were involved with all the tasks of field work. The diary entries of both genders suggested mutual respect for the contribution both made "for the good of the family farm."

Osterud's explanation for gender diversity resembles Adams's: "The allocation of particular tasks changed from day to day, depending on what else had to be done; from season to season, depending on the amount of milk being produced."[50] Just as Adams concluded, the range of tasks and the available labor pool meant that gender specialization would be irrational and inefficient.

These studies suggest that material factors, including crops, animals, and environmental conditions, influenced the gendered divisions of labor. There are not enough case studies to support a hypothesis relating farm diversification to patriarchal gender roles, but enough to suggest a connection.

A handful of scholars have examined how culture affects the role of gender in a rural division of labor. Sonya Salamon's study of ethnicity in midwestern family farming, for example, explores how the intricacies of family farm life—from child rearing to patterns of inheritance—are influenced by ethnic difference. Her typology, based on (German) "yeomen" and (Yankee) "entrepreneurs," shows that within contiguous communities, the customs and belief structures led to contrasting allocation of gender roles and farming practices. Salamon writes that "Yeoman and entrepreneur couples share a similar gender ranking of men over women, structurally resembling farm families in the past, . . . that does not . . . translate into conjugal farming teams organized identically."[51]

The contrasts between German yeoman and Yankee entrepreneur gender-based divisions of labor are not entirely consistent. The entrepreneurial farm family is more egalitarian than hierarchical in conjugal roles, but on the yeoman farm the women are more involved in production than on the entrepreneurial farm.[52] For my review the details of the distinction are less important than the realization that cultural features work their way into the structure of farming to define gender roles to the extent they do.

Osterud adds an interesting perspective on how culture may explain different gender roles on farms. As farms became commercialized, women withdrew from, or were forced out of, the dairy process (as pointed out by McMurry previously). Osterud reports that "by mid-century only hired girls [in New England] went to the barn; it was regarded as unladylike for a woman to wade through the mud and manure of the farmyard, to tuck up her skirts while milking, to consort freely with the men in the barn, and to carry heavy milk pails back to the buttery. When the dairy operation became a substantial source of farm income in New England . . . it came under masculine control."[53] In farm regions to the west, women participated more fully. Osterud writes, "Although the causes for the divergent paths of New England and mid-Atlantic farm women are not entirely clear, the consequences are striking. In New England, the gender division of labor on dairy farms was as sharp and rigid as on arable farms, while in the mid-Atlantic region women and men shared both the responsibility for and the labor of dairying."[54]

She makes the reasonable suggestion that differences in the gendered divisions were due to ethnic differences. Both male and female Scandinavian immigrants were accustomed to cattle work in their home countries, and the relative gender equality was preserved on early American dairy farms. Osterud writes that "Farm families in Wisconsin and Minnesota flexibly divided their labor within the context of joint responsibility for the dairy operation."[55]

These case studies suggest that material and cultural factors combine in any given system to create a division of labor with specific gender differences. In the background is a patriarchal society, which becomes more pronounced during certain historical eras and in the context of certain material systems. While certain case studies emphasize the importance of different elements, it is most useful to see their contributions to understanding a given situation, rather than to argue for a single mode of explanation which would cover all cases. In my case, patriarchy, material factors and culture are all part of the evolving relationships between domestic and productive technologies.

The Machines of Gender

Table 1 shows the evolution of productive technology on dairy farms. In concentrating on agricultural production, it tells one side of the tale. But each era is a result of both domestic and productive technologies, and these must be seen in relationship with each other.

On pioneer farms, men and women shared many of the jobs driving the productive technologies, while women largely monopolized the "domestic technologies." Women prepared and stored food, made cloth and clothes, and manufactured daily necessities such as soap and candles. Women used "domestic technologies" to care for minor maladies, to assist in childbirth, and to prepare bodies for burial.[56] While the elements in these technologies may have sometimes been simple, the working knowledge needed to make use of them was complex. For example, an individual might use hand tools to dig roots that would be dried, mixed, and otherwise processed into medicines. Because the domestic technologies were largely monopolized by women, they brought social power.

During the second era identified in table 1, women took responsibility for certain aspects of the productive technology. As McMurry described, women were responsible for household cheese production, and as Osterud notes, butter making was done by women in many New York dairy farms. Women largely took care of calves, poultry, and gardens and managed other agricultural processes located close to the house. As the productivity of these farms increased, however, women were replaced by hired men or by neighborhood labor crews. When women left the productive activities, however, the load borne by the domestic technology increased. Taking care of hired men and labor crews increased clothes washing by a quarter or more, when washing clothes meant pumping water by hand, heating it on a wood-

Fig. 89 Fair visitors inspect Gibson garden tractor, September 1947. Photograph by Sol Libsohn.

stove, and rubbing the dirt out of clothes on a washboard. And women were needed to prepare the large meals shared by hired men or the changing works crews.

The evolution of technology often brought the productive and domestic technologies into competition with each other. For example, farmers bought tractors before installing electricity and buying the domestic technology electricity would power. In practical terms, this meant, for example, that men would be pulling implements with a tractor across their fields as their wives were pumping water to hand wash their clothes on a board; cooking for family, hired help, or harvest crews with a wood-fired cookstove; and drying, pickling, canning, or otherwise preserving food, rather than simplifying these tasks with electric stoves and food preserved with freezing or refrigeration. While many farm wives struggled with these "technology

Fig. 90 Tired fair-goers, Bath fair, New York State, September 1945. Photograph by Charlotte Brooks.

wars," others accepted the logic that when the farm earned more money, it would be able to divert profits to improved domestic technology.[57] However, domestic technological improvements were often delayed for years and even decades.

The assignment of tasks on the basis of gender meant that the men's and women's identities evolved as gender roles changed. For example, the mechanization of agriculture removed women from meaningful productive roles. But during World War II, when many male farmers and most hired men were absorbed into the war effort, women were redefined as mechanically competent and responsible for previously male responsibilities. In fact, they had little trouble handling what were previously defined as "male" machines or tasks.[58] This paralleled the general Rosie the Riveter phenomenon whereby women moved aggressively into previously male roles in factories, foundries, and other hard-work labor. When the war was over, women were "reassigned" to domestic roles.[59] The domestic role, however, was in a process of redefinition as new household technology (modern kitchens and cleaning apparatuses) was adopted in rural as well as suburban or urban homes. Women were redefined as "homemakers"—and thus further removed from responsibility, decision making, and power on the farm.

Technological change, both in the productive and the domestic spheres, often had several unanticipated consequences. For example, when tractors replaced hired hands, it lessened women's domestic chores and removed an outsider from family life. Rural electrification brought the radio to American farmers, an expense which required about 10 percent of a typical farmer's annual profit.[60] In addition to liberating the farm families from the isolation they had long taken for granted, the radio also brought the ideologies of the national society to rural audiences.

These comments scratch the surface of the gender politics that surrounded the adoption of technology on dairy farms during the past one hundred years. Each of the changes, whether in the male-dominated productive sphere or the female-dominated domestic sphere, influenced the other. At the center was the question of how men and women sought power and gained social identities.

The Standard Oil of New Jersey Era

My analysis of gender roles during the SONJ era resembles Osterud's and Adams's portrait of patriarchy mitigated by gender flexibility inspired by evolving technology. Subsistence farming had evolved to rudimentary market production, which was on the verge of industrial intensification. These changes all had the effect of limiting women's productive roles. Finally, during the SONJ era, women began to be defined as needful of protection from the coarseness and physical demands of agriculture. The effect of this ideology was to reinforce the exclusion of women's participation in productive roles.

While the farmers I interviewed had generally organized their farms around traditional gender roles, the sex roles were not rigid. Mary said simply, "We all had to work . . . everybody had to work. There was nobody who sat down!"

Many women were, as a matter of course, integrated into productive agriculture. In the preceding chapter I described how Buck's mother, Gladys, a widow, farmed for several decades. The neighborhood admired her. Many other women assumed central roles on farms. Marian's example illustrates this well:

> *Marian:* I milked cows for twenty-nine years here. Some women are not involved at all—some do the book work, and then some do the same work as the men, and some milk . . .

Doug: Other people have told me that you used to work right alongside with Carmen.

Carmen: She used to help milk . . .

Marian: I did too . . .

Carmen: She used to help milk . . . about the only thing you ever done was you helped with the chores in the morning and the milk; the morning chores, like before we feed the cows and we done the milking and that was it. Then at night she'd go back out again, but you never had to do any of the cleaning of the barn, or that sort of stuff.

Marian: But I baled hay for many years. During one year, I baled all the hay.

Carmen: Yes, in the summertime, whenever we got the hay in, she always helped with the hay.

Marian: And I helped you with the grain . . .

Carmen: When?

Marian: Remember riding in the back of the binder . . .

Carmen: With the guy on the tractor and that old grain binder, yes, I guess you did—a few times— but that only lasted one year.

Well, she done her share out in the barn, no question about that.

Marian: And I had all those little ones . . . [their ten children]

Carmen: But some of them, some of the women, hell, they used to go out to the barn after breakfast to do the chores, but you never had to do that, so far as doing the chores or washing the milkers; I always did that.

Marian: I always washed the milking parlor down in the morning.

Carmen: Well, we didn't always have a milking parlor. Then she helped nights with the milking and haying, and helped nights with the chores. But cleaning the barn and all that stuff, why some of the women—they still do go out, not a lot of them, but I would say about half of them maybe, and not so much right around here, but in a lot of places they are working there, feeding the cows and all. But the kids were all small and we had them to look after. We'd come in and they would be up ramming around and getting into hell-raising.

Marian: We were up earlier, because I always had to get back in to get them ready for school. But once they got into junior high, they were almost always gone by the time I got in.

In this rather extraordinary passage, we see that Marian assumed central responsibility for farm tasks associated with animals and crops. She did this as she raised a huge family. At the date of the interview, their youngest child was fourteen years old, and Marian no longer worked in the barn or in the fields. At that time, following a pattern of many farms, she had begun a household business. By 1999, as a widowed and "retired" farm wife, she continued to run a bed and breakfast across the road from the farm now run by her son Kevin.

Fig. 91 Threshing wheat at the Beaujon Farm, Endicott, New York, August 1945. Photograph by Charlotte Brooks.

It is Carmen's grudging acceptance of Marian's contributions that are especially noteworthy in this exchange. Marian has had to argue for a history that includes her contributions. Carmen is able to remember when prodded, and yet on each account Marian must make her case anew. Given Carmen's characteristic decency and generosity of spirit, his inability to remember his wife's contribution to the farm is most telling. Curtis Stadtfeld's memory suffered from a similar sexist dysfunction. Upon examining the financial records of the farm on which he had grown up during the 1930s he commented, "as I looked through the books, something appeared that absolutely astonished me. Income from the eggs was often equal to or in excess of the income from the dairy herd. Yet the hens were always somehow regarded as a nuisance, kept by Mother for 'pin money' but not really a part of the farm system or the family economy."[61]

The flexibility of gender roles was evidenced in the way women and men shared poultry and egg projects on North Country farms. While it was generally a female task, both women and men remembered aspects of the work. I asked Marian if she had taken care of the chickens, and Carmen interrupted:

> She did the nice part of it. The collecting part!
>
> *Marian [interjecting]:* Well, we just had a few hens for awhile, but then during the Seaway construction, we extended the flock to about five hundred layers. We would get them as day-old chickens, and raise them until they started to lay, four or four and a half months.
>
> *Carmen:* We used to get five hundred hens, and then usually in the fall we usually sold a hundred or so of the pullets when they started to lay—we got pretty good for those because we got them early and they started laying in June, or July. We usually got them in January so that they were laying when the price of eggs was good. After you got warm weather they wouldn't give you over thirty cents a dozen for them, at that time. They were just a little extra money . . . added to what I got for working out with the chopper and the combine.

On Mary and Virgil's farm, the chickens were women's work:

> *Mary:* Now the oats are gone and I miss that. When we had the hens, you know, I always had the straw for my chickens. It'd smell so good when you'd go in there after cleaning them all out! It had a good smell to it!
>
> We had chickens until just a couple of years ago. We had chickens right up until four years ago. We'd have over a hundred. They were easy to raise . . .

But Mason also remembered raising chickens:

> I raised a bunch of Cornish giants, though, one year up there when we lived up there on the farm. I bought twenty-five chickens and I raised 'em. Raised twenty-two out of the twenty-five. Some of them dressed twelve pounds apiece. They get so heavy they couldn't walk. They'd just waddle all the time. Yeah, their feet'd all sprawl out like that.
>
> *Willie:* Just waddle.
>
> *Mason:* Yes. And the fat run right out of 'em when you went to dress 'em. Fat run right out of 'em. Of course, I fed 'em good. Oh, Jesus Christ, talk about good eatin'!

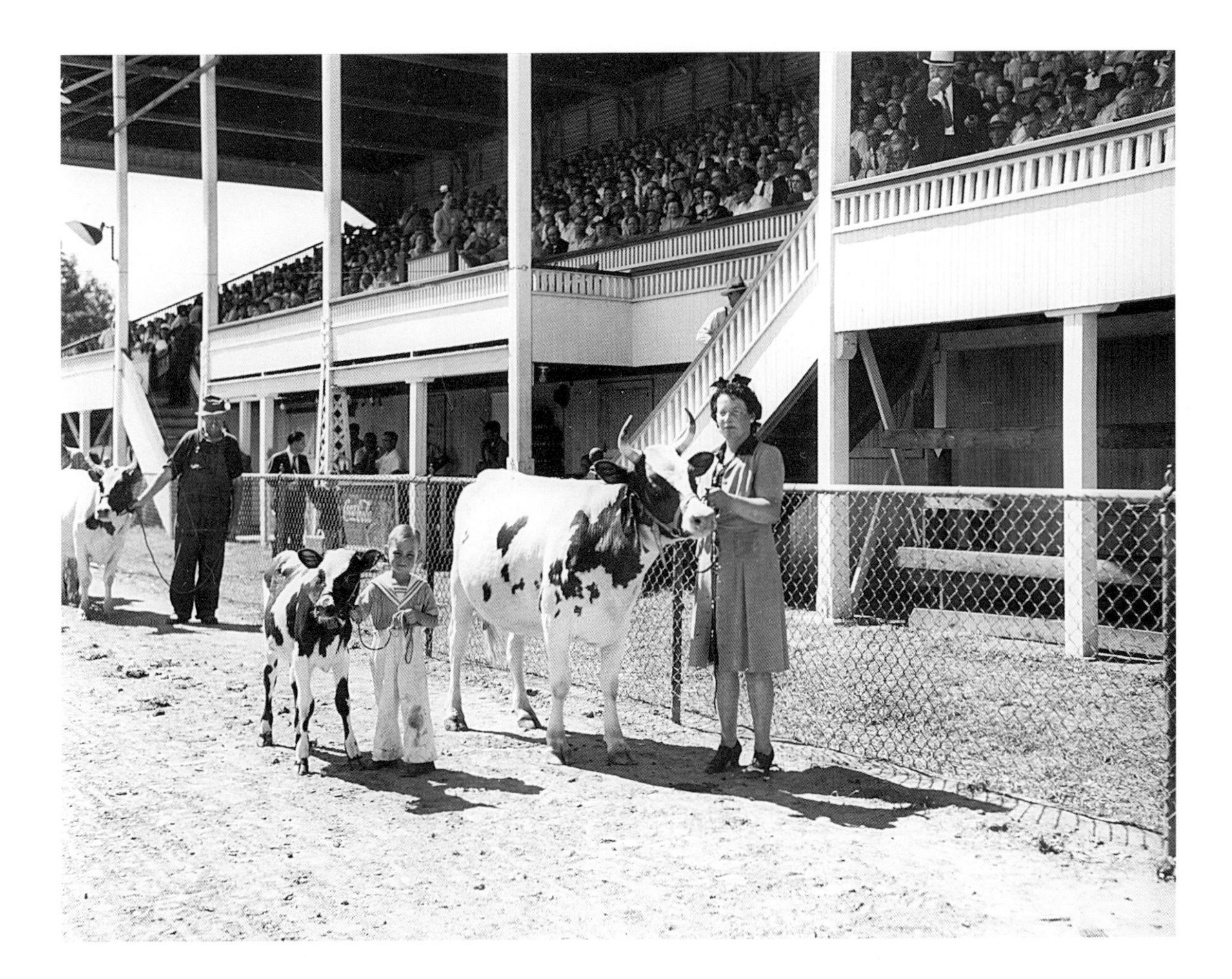

Fig. 92 Stock parade, Bath fair, September 1945. Photograph by Charlotte Brooks.

And Willie remembers butchering, which was generally a task shared by men and women:

> I've seen a time when we was raisin' chickens, you'd do about five hundred chickens a day. Just about do in your fingers!

These comments reinforce the sense of women and men sharing what in other farm systems was described as women's work. Several of the SONJ photographs, however, show women in "separate spheres" of "female" jobs and domestic chores. This included taking care of calves.

I asked Marie Lee if she was responsible for the young stock.

> *Marie:* I do all the calf work, yeah.
> *Doug:* Including helping with birth?
> *Marie:* No. I used to do that but I don't like it.

Doug: I wonder why the women always do the calf work? Seems like the same on every farm . . .

Marie: Because we treat 'em like babies. And they grow better, you know, if you give 'em that little extra attention and fuss over 'em a little bit and tell 'em they look nice!

Some women described pleasures that went along with their participation in productive tasks. Mary balances her dipping of the cream against her contribution as a worker:

> Did you ever eat cream toast? Real cream toast? Well, we do—we did. I said "just as long as I had to milk, I was going to have real cream toast when I wanted it." I'd go out and take it off the top of the can. He'd say, "Now you can't do that! Because it'd touch the test." I said, "Oh sugar! It won't touch the test at all!" It never moved it. When I'd go out I'd get a whole pint of nice cream—good cream. I wasn't cheating anybody—they knew I did it—it was mine; they hadn't bought it yet! So I'd bring that in and make it just like a cream sauce. Oh, was that ever good on toast! I'd heat a little bit of butter—we had butter, too—I'd heat it in the frying pan, and then I'd pour the cream into it, and just put enough flour in it to thicken it, so it would pour nice on your toast—salt and pepper and that was it. It was delicious! You can make milk gravy, but it wasn't cream gravy! [*she paused*] Oh, I miss that part; I miss everything that we grew for ourselves.

Ideology

Much of the discussion of identity and farm roles focuses on how women were defined into productive roles during World War II, and into the "maiden of the home" after the war. Jellison's research, cited above, is especially informative. But there was a more general ideological reorientation that accompanied the mechanization of agriculture. As Osterud noted, in English immigrant's eyes, it came to be seen as "unladylike" for women to milk cows, muck around in manure, be present during cow mating or calf birthing, or generally to be associated with men as dairy farming increased in scale. It is to be remembered that during the pioneer farm stage women were considered particularly adept at taking care of cows, precisely because of their alleged sensitivity to animals. In the following passage, Jim and Emily Fisher reflect on the shift in farm women's identities as part of a general social improvement of the rural life. The case of the Fishers is notable because Emily had not come from a farm background when she married Jim Fisher in 1936. Their farm was one of the most prosperous in the neighborhood, employing hired men and, often, a hired family, which made it unnecessary for Emily to contribute to field work. In addition, Emily was college educated and not drawn

Fig. 93 Judging canned gods at food exhibition, Bath fair, September 1945. Photograph by Charlotte Brooks.

to farm work by orientation. Her role on the farm, as supported by Jim, was auxiliary to the productive efforts:

Doug: What has happened to women's roles through all the changes recorded in these photographs?

Emily: The farm wife is just going right along with her husband, instead of helping drive the teams, she's probably driving a tractor when he needs her. But I shouldn't talk about women's roles on the farm so much because I never really did work like a great many women did. I never helped with milking and I've driven the tractor but once in a while, upon request and need. But I've never done a lot of the hard work that a lot of people have done just as a matter of course. Those women were expected to do milkings and take turns, and do the work right along with their husbands. But Jim had people to help and I had to board the people who helped, so you know, I had that to do. And I did all of that gladly, and then we had children and a number of older relatives, so I've never been the out-at-the-barn type person. Not that I disliked it—I enjoyed the farm, but I'm not a worker, like some of them have been.

Jim: Well, I was trying to think about it. I would say the women in this community have never been real farmers in the sense of the word that you think about women getting out and shouldering most of the responsibility.

Emily: I've run a lot of errands.

Jim: Oh, goodness yes!

Emily: And driven a lot of children here and there.

Jim: Yes, homemakers. I see that some single women now even choose farming as an occupation, and it's not impossible for them to do that by gearing up with some of these modern conveniences, like the pipeline, and keeping it to a small farm that they can manage efficiently. It will be interesting to see how some of these farms work. My mother never worked. My grandparents were more like the pioneers, where there was hand milking. I remember my dad saying that his mother took him to a barn and propped him on a board that was strapped against the wall while she milked; that sort of thing. There were single women in the household in those years, because there were a couple of maiden aunts that lived there and helped at the barn with what wasn't heavy work.

But then the next generation sort of took a little different view of women working at the barn. They had all been to college, and my grandfather and my father just sort of took the attitude that it coarsened a woman to work out like a man, and they liked them more ladylike. The woman's place was in the home, and they liked a nicely established home, and you couldn't have both. If you had someone working outside all the time the housework was neglected, and the child raising was neglected. I'm sure that was the philosophy back in their generation. And there was some merit to it. You've seem some women who work out all the time and it does tend to change them from a feminine point of view, or from being the best kind of homemaker. That was the attitude my parents had. In fact when Diana, our older daughter, joined 4-H and was showing cattle, Dad was always right around, seeing to it that she wasn't doing the heavy, dirty work, you know. He felt that it coarsened a girl, that she shouldn't be pitching the manure and right in there doing the rough jobs like the men should be doing. A little bit of chivalry there, and I'm sure they all felt that. You know, I'm always a little bit amazed yet to see a girl right in there pitching shit, getting so crummy and dirty looking. I don't know!

I mention that my wife cleans the barns for her four horses.

> *Jim replies:* Oh, it's different with horses, there's not that awful round of filthy work. Horse manure is dry and easy to lift. I don't feel the same way about that at all as you do with a bunch of dairy cows. It's such a continual round of keeping clean!

Jim and Emily accepted gender-based sex roles as an aspect of the evolution of their farm to scientific, rational, and machine-based. But Jim's memory of his father's protection of his daughter from work which would "coarsen" her identifies an ideology in which men define the proper role for women while removing them from central responsibility of running the farm. The relative gender equality of the first farms came to be defined as lowering women to the level of men, working with animal excrement, heavy tools and machinery, and sweaty backs. To be truly a woman, they were required to move into roles with less social power.

The Gendered Gaze

How do we see gender roles in the SONJ photographs included in this volume? There are two parts to this inquiry, the first to note what men and women were doing on dairy farms during the World War II era, and the second to analyze how male and female photographers portrayed this reality. My point is that the visual record is partly constructed through a "lens of gender"; the photos are true in the sense of recording something that appeared in front of the lens, but they are also a function of the photographer's choices. Perhaps male and female documentary photographers found different subjects worthy of documentation; thus, perhaps what we might mistake as an objective history can be better understood as a construction based on how women and men saw the worlds they photographed.[62]

In the first stage of photo research I selected about two hundred photographs from the SONJ archives by looking for images that showed objects or events I'd heard dairy farmers describe. I was not, at first, aware that this selection of photographs included many by two SONJ photographers, Charlotte Brooks and Ester Bubley, who were women, since the archives list only the last names of the photographers.

In a second edit, I selected about 125 images, which I further reduced to about 100 images included in this manuscript. The following comments refer to the approximately 125 images which were the core photographs around which I organized this project, both the published photos and some which were in the running until the final edit.

I did not analyze all of the SONJ images relating to dairy farming as to the role of gender in defining work on the dairy farm. A thorough study of the topics pho-

Fig. 94 Signs of autumn: harvest festival, Methodist Episcopal church, Newark Valley, November 1945. Photograph by Charlotte Brooks.

tographed by SONJ archives would offer a way to look at how men and women photographed several subjects, since the professional charge given to each photographer, regardless of their sex, was the same. For my study of dairy farmers, an analysis of the photos I did use may tell us something about how the photographers saw social reality through the lens of gender.

Forty-one of the approximately 125 images I selected for this project were taken by Charlotte Brooks, the only woman represented in the selection from which I assembled this project. As I consider Brooks's photographs to be examples of the female "gendered gaze," I am, of course, allowing a single photographer to stand for women as a class of photographers. This is not an acceptable basis for a firm conclusion. It would be easy to discount any pattern I discover as a result of individual working habits or idiosyncrasies. Yet as I looked at these images from the perspective of gender authorship, I found striking patterns. Whether they are due more to Brooks the individual or to Brooks the woman cannot be known.

Many of Charlotte Brooks's images were of shared work and details of complex jobs. These include figures 56 (a threshing crew setting up), 73 (a crew gathering corn in a field), 75 (chopping corn at a silo), 76 (inside a silo), 79 (cutting wood together), and 83 (a long shot of threshing beans in a field). Male photographers recorded some of these views, such as Libsohn's photographs of the details of grain threshing, but Brooks seemed particularly focused on detailed images of collective work. As I compared Brooks's images to those taken by men, it seemed that men more often emphasized objects, whereas Brooks more often recorded action. But the difference in these styles of portrayal is not overwhelming.

It is possible that a woman's point of view sensitized her to work as constituent elements, or maybe it was just that she, among the SONJ photographers assigned to these subjects, happened to see social life as organized around work which was best understood as series of steps, often shared in a teamwork arrangement.

More telling is that of the forty-one images by Brooks that I used in my project, nine include women as a primary subject. That just under 25 percent of Brooks's images focus on women is made more striking by the fact that only three photos of the approximately eighty-five remaining images in my collection, all taken by men, include women as a primary subject. Men's photos of women include Gordon Parks's photo of a woman stooking corn, the only photo of women working in an agricultural job (fig. 71); Sol Libsohn's photo of a woman serving dinner to a changing works crew (fig. 82), which was, in fact, the second photo of a series of two showing the changing works crew eating dinner (the first is fig. 80); and Libsohn's photo of women at the auction, also part of a pair of images showing both men and women at the same function, which appears as figure 86. It is incredible to note that men almost never photographed women, and when they did, it was typical to portray them as auxiliaries to men's worlds.

Brooks's images of women recorded food preparation (figs. 81, 87), the participation of women in the harvest (fig. 91), and women's' involvement in the agricultural fairs (figs. 92, 93, 99). Noteworthy in figure 92, her subject is a mother (herself leading a cow) and a young boy, probably six or seven, leading his calf in front of a viewing stand. The boy, dressed to the nines, has a sailor suit on, but a bit messed up from the work of keeping track, one assumes, of the calf. The mother is present for her child, helping him through an important moment. Similarly, Brooks's image of teenage boys and girls in a church-sponsored harvest celebration (fig. 94) is one of the few in the entire collection that deals with the socialization of young people. Here they are well scrubbed and enthusiastic in what is probably a celebration of a shipment of donated food (none of my subjects could guess exactly to where!). Sol Libsohn's photos of the state fair (figs. 21, 89, and several not reproduced in this volume) show men gathered around machines, often observing them in demonstrations. Brooks's gentle image of men at a fair (fig. 90) shows them weary, perhaps vulnerable.

Fig. 95 Corner of the summer kitchen where popcorn has been hung to dry, Brooklea Farm, Kanona, New York, November 1945. Photograph by Charlotte Brooks.

But while these images help us understand the role of gender on the farm and in defining the small farm, they are an incomplete portrait of women's roles in the dairy farm culture.

Notable by their absence are images of such topics as women's full role in the productive end of the farm (women typically took care of many animals, including birthing and caring for calves). Women are not shown in their full roles as housekeepers (there are few photographs of routine housework; fire tending; cleaning, washing, repairing, or making clothes; or child care and family maintenance through recreational and educational work with children). Finally, there are few if any images which explore how women's hidden work maintained the emotional and psychological aspects of family life.

The SONJ archives portray dairy farming in this era primarily from a male perspective. We can feel fortunate that the work of one female, Charlotte Brooks, partially addressed this imbalance. Brooks's images help complete the picture, but she could have gone much further. If we read these images as history, we must see this history as constructed around the power and blinders of gender.

Souping Up Cows

The mechanization of dairy farming during the SONJ era redefined the relationships between cows and farmers and, in the process, changed the cow. Once a partner in a decade-long relationship with a farmer, the cow became, for most farmers, a souped-up milk machine, to be run hard and slaughtered early. This transformation continues in several different ways, as described in the final chapter.

My intent is to show the evolution of feeding, breeding, milking, and housing cows. I also describe how, in the earlier eras, institutions such as 4-H passed these practices to new generations of farmers.

At the center of these dramas, of course, remains the cow. Curtis Stadtfeld, the child of a dairy farmer, wrote: "Cattle have more personality than is usually recognized, and in those days of small herds there was a strong bond between the owner and his cattle. Partly it grew out of memories of the times when men and cows needed each other so desperately, when they served as beasts of burden as well as giving a little milk to keep the children healthy; partly it was because they are fully domesticated, really unable to fend for themselves, and as such are, like children, a burden that fills for the farmer some of the need to be needed."[63] He elaborated: "Care of a dairy herd is an endless task and can take as much time as the herdsman wants to spend. But it develops a healthy affection between herdsmen and herd."[64]

This said by way of introduction, cows are essentially milk machines. The necessary ingredient is sperm, in the past deposited by a bull, a beast of uncertain temperament and great destructive power. Natural rhythms coordinate a cow's pregnancy with the cycles of nature:

Fig. 96 Leroy McKenzie bringing the cows down from the hills in the evening, Shuman Bros. Farm, Roxbury, New York, August–September 1945. Photograph by Sol Libsohn.

Tom Lee: A farmer would aim to breed about May and June, when they put their cows out to pasture. The nutrients in the grasses would rejuvenate their reproductive systems to they'd come in heat better and that's the natural time for animals to breed . . . there's a lot of factors. Their peak time of production is about two months after they calve. And late May and early June is when the peak amount and the peak quality forage is out there, so that's the way Mother Nature would do it. If your calves are born now she can start feeding them at this time, when she doesn't need to produce a lot of milk. In a couple of months the calves are going to be bigger, and going to need more milk, and the grasses are going to be there, so that's about the time a bull would naturally breed 'em. And in those days, that was quite often the way. My father often told me that in January, February lots of farmers did not milk any, or just one or two cows—just enough for house milk. And the rest of 'em would be dried off.

The records of milk checks from a twenty-two-head herd in 1939, shown in figure 97, show the ebb and flow of milk production over a year. Several farmers told me how, in days before artificial insemination, these natural cycles produced a boon in milk when the demand was lowest and a bust in milk production during winter months, when urban schools created high demand. So the price for milk during this era substantially increased in the winter, when the farmer had little milk to sell. The other side of this picture is that the farmers and the cows rested together in the winter. The year's seasonal rhythm lessened the winter work and demanded the most in fairer weather, when work was easiest.

To be a farmer on the most rudimentary level required only a few dairy cows, the services of a bull, and naturally growing grasses or other feeds. The ability of cattle to subsist on a wide range of foods, and to turn these various substances to milk, makes them adaptable to a wide range of environments.

The cow's nine-month gestation produces milk nearly year-round, and a cow can have a productive life of fifteen or more years. When the cow bears a female, she has made a new-generation milk machine. If the calf is born a bull, it is raised briefly and then slaughtered for veal. Now and then, a farmer needs a new genetic stock or a younger bull, and the animal gains a few seasons of cow servicing before becoming hamburger.

At its core, it is an udderly simple system. Cows are creatures of habit and usually stay healthy; they can usually be tricked into thinking they are not equal in strength to their human masters. Bulls, of course, are another matter; they are volatile and unpredictable. But if the animals are allowed to roam, if nature runs her course, pregnancies repeat and the cows produce milk.

Feeding Cows

Dutch and English settlers established dairy farms during the seventeenth and eighteenth centuries in what became New York and surrounding regions. Cattle farming was well developed in Holland at that time, and settlers brought sophisticated farming practices with them to the Americas. As farming extended into the frontier, however, they became more rudimentary. David Cohen quotes the Swedish naturalist Peter Kalm, writing in 1748, that "cattle had no stables in winter or summer and were obliged to graze in the open air during the whole year."[65] These early farms included many kinds of animals, including milk cows. A 1638 *Inventory of Farms in Eastern New York State* listed farms with between three and six milk cows.[66] In 1675, tax lists for towns on Long Island showed an average of 8.1 cattle per Dutch farm and 9.8 for English farms.[67] One can assume that this

12 Milk Checks 1939

	test	hauling	lbs		total
Jan	3.65	10.87	10876	1.86	202 29
Feb	3.5	~~12.069~~ 12.06	12,067	1.52	170 15
Mar	3.35	15.66	15,669	1.16	164 54
april	3.25	19.12	17,399	.87	132 25
may			19,091		163 68
June	3.4		22,728	1.04½	237 50
July	3.3		18.044	1.34	241 78
aug	3.4		15,569	1.8426	286 87
Sept	3.45		19,553	1.9833	387 79
Oct	3.4		18,091	2.1484	388 66
Nov	3.55		11624	2.2162	257 61
Dec	3.95		9,173	2.1764	199 64
			189,864		2882 36

Fig. 97 Records of milk checks from a North Country farm in 1939. These records show how the price of milk fluctuated inversely with the quantity of milk produced. Given the natural cycles, the amount of milk produced on a twenty-two-head herd doubled between January and June. The second column from the left, labeled "test," indicates the butterfat for the milk. Farmers were paid a premium for higher butterfat. The next column records hauling costs, a sizable percentage of total income. The fourth column indicates the price paid per hundredweight, and the final column is the total of their milk checks.

Fig. 98 Dairy farm on Route 163 near Fort Plain, New York, August–September 1945. Photograph by Sol Libsohn.

included about six milkers per farm, the beginning of the dairy system that was to flourish during the nineteenth century in northeastern states.

On early farms, cows mostly fended for themselves during summer months, grazing through rough pasture, woods, and swamps looking for food. Because farmers were not, at first, efficient crop growers, there was scant hay in the long months between the growing seasons. It was not uncommon to slaughter cattle in the fall when farmers anticipated that winter feed would fall short.

By the mid–nineteenth century, a typical dairy farm included land both to pasture animals and to raise hay in meadows which would be harvested and stored for winter feeding. In lush northern dairy regions, a cow needed two acres of pasture for summer feeding. Native mixed grasses made excellent feed and replenished themselves naturally. As the system evolved, farmers replaced native mixed grasses with single species. Some species, such as clover, had nitrogen-fixing capabilities;

thus, the crops were partly self-fertilizing. In mid-nineteenth-century in New York, it was customary to add locally mined lime to adjust the soil pH. As barns were developed for winter housing, it became easier to gather winter manure, which farmers used to increase soil fertility.

With postwar mechanization, herd sizes increased and cropping practices rapidly changed. To maximize crop output, farmers plowed and planted pastures and began keeping cows close to the barns to be fed from harvested crops and purchased additives. As farmers replaced horses with tractors, there was less need for oats and barley. Farmers began to purchase protein supplements to enhance their cows' diets. As we have noted, farmers who had grown grain individually but harvested it collectively suddenly found better use for the land and their time. Within a very short period following the SONJ era, northeastern dairy farmers stopped growing small grains altogether, and pasture was largely eliminated.

These feeding practices had several effects. "Balancing" the rations of the cows made them more efficient milk producers. Keeping cows from the pasture decreased the animals' exercise and placed them in crowded, manury, and less healthy settings. The use of more and more fermented feed (including fermented hay) required the addition of ingredients such as sodium bicarbonate to the cows' diets. The cost of feed additives, including those listed and several others, dramatically increased the costs and complexity of dairy farming.

The biggest single transition was a severalfold increase in corn production. Corn produced a greater tonnage per acre than any other crop, and animals could eat corn stalks as well as the kernels. It seemed a natural transition to what seemed to be a better source of animal feed.

The downside, however, was that corn is a "heavy feeder"—it rapidly depletes soil nutrients and, therefore requires heavy infusions of fertilizer. In addition, as farmers moved to larger herds and two crops, the simplified ecosystem made these crops more vulnerable to pests and disease. The chemical response in pesticides and herbicides shifted the farm further from its previously natural cycles.

We see in figure 99 the shocking image of the pesticide DDT (developed during germ weapon research during World War II) being sold at the state fair: a lumpy salesman hands a farm wife a four- or five-dollar pint-sized container.

Arnold Eagle's photograph of crop dusting in Maine (fig. 100) shows another World War II invention, the helicopter, applied to newly industrialized agriculture. In this chilling photograph, a man crouches in the field, about to be covered by poison. John Vachon's image of a "crop duster" documents the lack of caution with which these poisons were handled and dumped into the environment. The plane veers over a rural road, climbing to avoid power lines and cars

Fig. 99 DDT display at fair, Bath fair, September 1945. Photograph by Charlotte Brooks.

driving past (we can imagine their windows open), the insecticide still spewing from the biplane.

While the aerial application of chemicals did not, in the long run, prove to be practical on the small farms of the northeast, it became typical to rely on inorganic fertilizers, pesticides and herbicides.

Herd averages (the average yearly production of a cow herd) increased steadily from the mid–nineteenth century. McMurry argues that the increased milk yields between 1850 and 1900 were due to "the cow's ability to make use of extra feed,"[68] rather than to genetic improvement of the animals themselves.

Fig. 100 Aroostook potato country in the vicinity of Presque Isle, Maine: helicopter dusting a potato field against blight, August 1949. The Aroostook Agricultural Experimental Station reports a 10 percent increase in yield in helicopter-dusted fields because of the absence of wheel damage. A helicopter can dust six hundred acres a day. Photograph by Arnold Eagle.

North Country farmers had a long tradition of adding supplements to their homegrown feed:

Mason: We took the oats to get ground up in Lisbon, at the grist mill. They'd make up a ration for you, you know; they'd add other stuff like soybeans and bran and cornmeal to bring it up, like, to 18 percent [protein]. Oats is only like 10 or 11 percent.

Doug: So they were mixing feed even in the thirties?

Mason: Oh, yeah. Yeah.

Willie: We used to have trucks go around.

Mason: Grinders on 'em.

Willie: Regular grain grinder onto 'em and they'd mix it for you. Grind it and mix it.

Mason: Bob Fisher had one. Used to go all over with it.

Willie: Yeah, well, it ain't been too many years ago that he quit doin' that.

Mason: Like if you just had corn and oats you'd grind it together, you know. But you could feed cows corn and oats, you know, and get by with it.

The transformation of farms is reflected in how cows were fed. In the original system, a farmer fed each cow individually:

Virgil: The cows would come in and go to their stanchions. They knew

Fig. 101 Seabrook Farms, near Bridgeton, New Jersey, August 1947. Four planes are used during the growing season at Seabrook Farms to spray insecticide on the crops. One plane can cover fifty acres an hour. Spraying must be done at sunrise before the ground gets too hot. Photograph by John Vachon.

where they belonged. If they went to the wrong one we'd move them back to where they were supposed to be. We'd start them with their calves . . . we kept the calves right in the barn, and then in the spring we had a calf pasture. We'd feed them right there. I'd give each cow what she needed, depending on how well she was producing. I knew each cow.

Mason explained the feeding routine:

Mason: We'd feed the grain before you'd milk.
Doug: A couple of scoops for each one?
Mason: Yeah. It all depends on how much milk they give. You feed 'em according to what you give 'em for milk. Like, if they give twenty pounds of milk, why you give 'em five pounds of grain. So it's one to four. That's the way we used to do it anyways.

Early farmers fed cattle poorly and allowed them only scraggly grazing land. As a result, the animals suffered and the farmer realized little of their potential. They also abused the land with overgrazing and little crop rotation or natural fertilizers. But farmers eventually learned that land, feed, and animals could work in an interlocking harmony. A quarter section or two could nearly supply the feed for a herd large enough to support a family. With tools which were not extraordinarily expensive, farmers could grow and harvest crops which fed their cows well, if not perfectly. Rotation of crops is more complicated than replanting King Corn year after year, but rotation lessens the chemical addiction of the modern farmer. In fact, most farmers do not want to be chemical junkies but envision few alternatives, even though alternatives exist even in their memories. Finally, one might ask the responsibility of the farmer to the cow. Is there not a place in the argument which addresses how cropping and feeding practices affect the animal? Is the image of the cow in the pasture (fig. 98) not only about our pastoral stereotypes but also an image which suggests the right of the animal to a reasonable existence?

Fig. 102 Measuring feed, dairy farm, Maple Lane Farm, Greenport, October 1945. Feed for registered Guernseys is measured at each feeding. Photograph by Gene Badger.

Breeding

Thousands upon thousands of breeding decisions and environmental conditions have produced, over the centuries, animals with different characteristics. Centuries before Darwin, farmers were aware of the principles of genetic variation and resulting adaptability of species. They noted characteristics of various animals and arranged breeding so that those with the most desirable characteristics provided offspring.

What is "desirable," however, varies from place to place and from one historical era to another.

For example, farmers might seek cows that live easily in the cold and thrive on spare local feed. In regions where feed is plentiful, farmers might breed to enhance milk production. In regions where people develop tastes for specific cheeses, cows will be bred, over time, to produce milk with certain butterfat or protein levels, which produce those cheeses. For example, Jersey cows, developed on the Island of Jersey, are small, hardy cows (typically weighing less than one thousand pounds) which produce milk with a butterfat content as high as 6.5 percent. They are also friendly and thought to be smarter than others. In an era in which feed was relatively harder to grow and butterfat was looked upon favorably, Jerseys were a breed of choice. They were, for several decades, the elite cows of the North Country dairy farms. One or two farms in my study still retain exclusively Jersey herds.

Holsteins, the breed that largely replaced Jerseys (and other breeds), weigh one-third more and produce at least one-third more milk, but with half the butterfat content. They are high producers, big eaters of what old-fashioned farmers call "diluted" milk, and not known for their personalities. They are, indeed, the milk machines of the modern dairy farm. They have been bred for even greater productivity, and cows on leading farms now produce more than a third more than they did during the SONJ era.

This discussion, however, makes breeding sound simple. In reality it is not. McMurry indicates that pre-nineteenth-century diary herds in the United States were a "motley crew" of accidentally mixed European cattle, and that

> the art and science of breeding was a mystery even to the successful. . . . Given the paucity of knowledge about heredity, writers acknowledged the unpredictability of attempts at systematic breeding. . . . It is not surprising that nineteenth century livestock owners were puzzled by the vagaries of breeding. Even today's sophisticated geneticists have not unraveled heredity in all its complexity, and it is still very difficult to sort out the differential effects of heredity and environment.[69]

Arthur put the issue in the concrete:

Doug: And you'd breed on the basis of what you'd learned from the previous generations?

Arthur: Don't talk to me about it! That's a story by itself. There ain't many got enough brains to do it. That's a lost art . . . they never learned all there was in terms of breeding!

I ask him to explain.

Arthur: Well, yeah, I'll have to tell you. A lot of times the son is never as good as the old man. Or the daughter is never as good as the old lady. You know what I mean?

Doug: Yeah?

Arthur: And then you go ahead and you have a cow that is best in the business . . . you find the bull that you think is the best, and it don't work. . . . It's the same way with the people. Did you follow me? Or am I wrong?

Doug: Well . . .

Arthur: I'm wrong?

Doug: I don't know. What's the point?

Arthur: The story is that it don't work always. You can mate a high butterfat cow with a bull that had a high butterfat mother . . . and you cross 'em and she may not go with the genes . . . Well, even science don't work. If it did every time I'd have professional cows.

Doug: How do you do it then?

Arthur: Well, you pick out the bull you want. You want more butterfat. They have the association, they have a bull with high butterfat and a black bull with four good feet. . . . So I'd like that bull. I like that bull. Well, that's the one I select. You may not like that bull because he's too black and you may want him to give more protein in the milk so you take another bull, maybe a white. So [just as] individual people are, every one of them, different [so are cows].

It was during the SONJ era that farmers, assisted by agricultural agents, began to systematically gather information which made breeding more scientific and focused more specifically on milk production. Charlotte Brooks photographed a step in this process (fig. 103).

Jim Fisher reflected:

Oh my goodness, that is an amazing photograph. In the early years when the DHIA [Dairy Herd Improvement Association] sampler came along, he took a sample of each cow's milk and put it in a little tube, and then he stayed overnight at the farm. Then the next day the samples were all run through a whirling bath of warm water called a centrifuge. It separated the butterfat

Fig. 103 Bruce Smith, Dairy Herd Improvement Association inspector, testing milk for butterfat content, Brooklea Farm, Kanoa, New York, September 1945. Photograph by Charlotte Brooks.

from the milk, and they had it all calibrated so that after it went through this centrifuge they added enough of some solution so that it floated the fat up into this little neck. Then with little tweezers or calibrators they read whether it was three percent, four percent or four five or whatever, and then that was written down in his book, and that was multiplied by the number of pounds the cow gave per day, and that was how much butterfat she produced for the day and then for the month. I had forgotten all about that. My goodness, yes! Phil Babcock was our cow tester who came for years—as a young man, I was awfully intrigued by the process.

Doug: He stayed overnight at the farm?

Jim: Oh yes, he took a sample of the night milking and the morning milk. Each cow's milk was weighed and stirred. A sample of milk was drawn and put into a flask—night and morning—and then it was all shook up drawn with a pipette—and put into this little bottle with a long slim neck, and then it was popped into this centrifuge. It was cranked by hand for so many minutes and then opened up. You could see that it was separated out, and there would be a little line of butterfat that looked like butter in this neck of the bottle. Then they sat down and read it all and took the number and recorded it on a page. Good heavens—I hadn't thought of that in years and years. But the efficient farmers did that, and that was a way of knowing what your cows were producing.

Doug: That would influence, then, the breeding decisions?

Jim: Oh yes, the cows that were producing the most milk per year, of course, were the ones you arranged to have the calves from. And if they weren't producing well, or were not producing much butterfat and not much milk, then they were sent off to slaughter—you would never think of raising a heifer calf from that!

Bulls and Queer Steers

Breeding naturally occurs during the spring and summer, so that the natural gestation produces calves when spring pasture is plentiful. Farmers traditionally sought to improve their herds by keeping a bull calf from a mother who had desirable characteristics and using that bull to impregnate other animals in the herd. A constant problem was inbreeding. To overcome inbreeding, farmers traded bulls or arranged for well-regarded bulls to visit their herd.

When it became possible to inseminate artificially, the genetic possibilities increased far beyond anything available in the previous system. Even more important for the transition to a more rationalized system, artificially inseminating cows during all months of the year meant that milk would flow at the end of all these gestations, now spread throughout the year. Breaking out of the cycle of natural breeding then had the unanticipated consequence of necessitating better-quality feed year-round.

These transitions took decades to complete. For many years, farmers both used artificial insemination for most cows and kept a bull for cows that did not breed easily (injecting many shots of semen into a cow that does not breed easily is expensive, but a bull provides stud service at essentially no additional cost to his normal boarding).

Virgil: We began to use artificial insemination when it first came in. We didn't keep our bulls after that. Before, we'd just put the cows out with the bull in the spring—all of them together. They'd get bred in about a month. We'd have a lot of milk in the summer [the year later], and in the winter we wouldn't have any! Some of the cheese factories would close down in the wintertime—there wasn't any milk! Our income went almost to zero. We had to plan! After we got the artificial insemination we spaced out the cows, and the milk production went up. The inseminator had a book with the records of the various animals: what the butterfat was . . . the production . . . you picked what you wanted.

I think the insemination cost six dollars when it started—it was expensive at that time. But if you figured what it cost to feed a bull all winter—all year—and you raised one up from a calf—it really wasn't so bad!

Mary: And you didn't have to be afraid to go over to the berry patch!

Virgil: I had one get me down in the barn, too. I wasn't sad to see the bulls go.

Jim Fisher describes an intermediate step:

If you were in doubt about when they aren't conceiving, you would breed them when they first came into heat, and maybe in the middle, and again at the end. They are a little more apt to conceive if you don't breed them too soon on the heat period—more as they are going off the heat. But of course they need to be standing if you are using a bull. If you are artificially breeding, and can put the sperm directly into the uterus, then you can breed later and get fairly good conception rate. From the stage of using the proven bulls, a number of us got to getting the artificial breeding equipment and drawing the semen from our own bull, storing it in the refrigerator, and trading that back and forth with the neighbors. We went through that process for quite a while.

Doug: Everybody learned it together?

Jim: Yes, there was a Jersey breeder in DePeyster who was interested in the breeding in our particular herd sire. He hired me to breed six cows one year, and I went up there with the semen and it just happened he got four heifer calves from those six cows. He was quite delighted!

The process Jim describes, collecting sperm from one's own bulls, includes stimulating bulls with "queer steers" and then collecting semen when the bull is unable to penetrate the steer. This process was not for the faint of heart.[70] Eventually, breeding specialists refined these tasks and became semen salesmen. It was a more efficient way to make milk, but with these changes an element of the old farm culture disappeared.

The SONJ photos of bulls evoked many memories:

Mason: We'd always buy a bull. You'd never keep a bull out of your own dairy; you didn't want 'em cross-bred . . . you'd keep a bull about three years.

Willie: —that was average, three years.

Doug: Did they get mean?

Mason: They'd run you right up the goddamn apple tree! But we had a cow dog, mister. All we had to do was whistle to the dog, mister, and the bull . . . he started lookin' . . .

Willie: He'd perk himself right up.

Mason: He was ready to head for the barn. Because when that dog come around the corner of the barn Mr. Bull was a-travelin'.

Willie: Yeah, we had one the same way.

Mason: Oh, he was scared to death of that dog. Oh, good cow dog. Every time she'd bite him he'd bleed. And she never would do that to a cow.

Willie: No.

Mason: But she hated that bull.

Willie: Any bull they hated.

Doug: Why?

Mason: I don't know why. That particular bull was treacherous. You could go out there to the pasture to get the cows to bring 'em in to put 'em in the stanchions and, of course, if the dog wasn't with you—

Willie: The bull didn't want 'em to get away.

Mason: —oh, that bull, he'd bellar; he'd paw and throw the dirt. You'd whistle to the dog—And he'd bellar. He didn't want the cows to come in, see. Whistle to the dog. The bull's ears'd perk up. . . . He'd stop—

Willie: —and he'd look . . . and start . . . see where the dog was.

Mason: See if the dog was comin'. When that dog come around the goddamn corner of the barn, Mr. Bull, he was a runnin'. Heh . . . heh. . . . And he had to come in the door and the box stall was there.

Willie: Miss it half the time.

Mason: . . . he'd hit that right wide open. Smack! And the dog would stand right there 'til you got the door shut. Mr. Bull never come back out. Best goddamn dog that I ever seen in my life!

Buck Henry, who trucks animals as well as running his dairy, saw a lot of his life in figures 105 and 106:

Buck: It don't look really any different today at an auction. The only thing I see different here is that they are usually fenced in so the animals are away from the crowd. Here he is not away from the crowd, you know. A bull like that can be dangerous—it all depends on his personality. I took bulls now and then—they can be real dangerous. There are still bulls around the country. I would say 70 percent of your farms have a bull on them.

Doug: Would you use a bull for the heifer for the first time?

Buck: For the hard cows. The ones that don't breed easy. They artificially

Fig. 104 S. B. Moore leads his bull out to the pasture for fresh air, August 1944. Photograph by Gordon Parks.

inseminate them twice; then if they don't stick, you use the bull. If they're not raising any calves, and they're buying their cows, they would just use the bull.

Doug: Why would that be?

Buck: They just don't want to pay the money for the semen because they can raise the bull or buy the bull—whatever they do—keep him two years and they got the bull to sell for probably half again more than they paid for him.

Fig. 105 Cattle trucks at Jay Madsen's livestock market, Bath, New York, September 1945. Photograph by Charlotte Brooks.

He is making them money growing as well as doing what they were paying the inseminator to do.

They just slaughter the calves anyway. They just have the bull there so they can have a calf. So they'll freshen . . .

The bull to me is bad news. I've got a Jersey bull in the field right now with the heifers. I had one in here in the spring I had to get out of here. He got ugly. Like, if you pulled up in the yard like you did tonight, he would come right over to the fence. He would scare you.

I got to where I would sit here in the house and I'd think "I hope nobody walks down the road." You have one little string of electric fence there. This guy's acting like a monster. That electric fence wouldn't mean diddly boo to him when they go out of their head. You get a bad one—he goes crazy—his eyes turn red. They just get so mad. They go completely berserk. They will run their tongue out; they'll make all kinds of squealing noises. I mean they don't bellow;

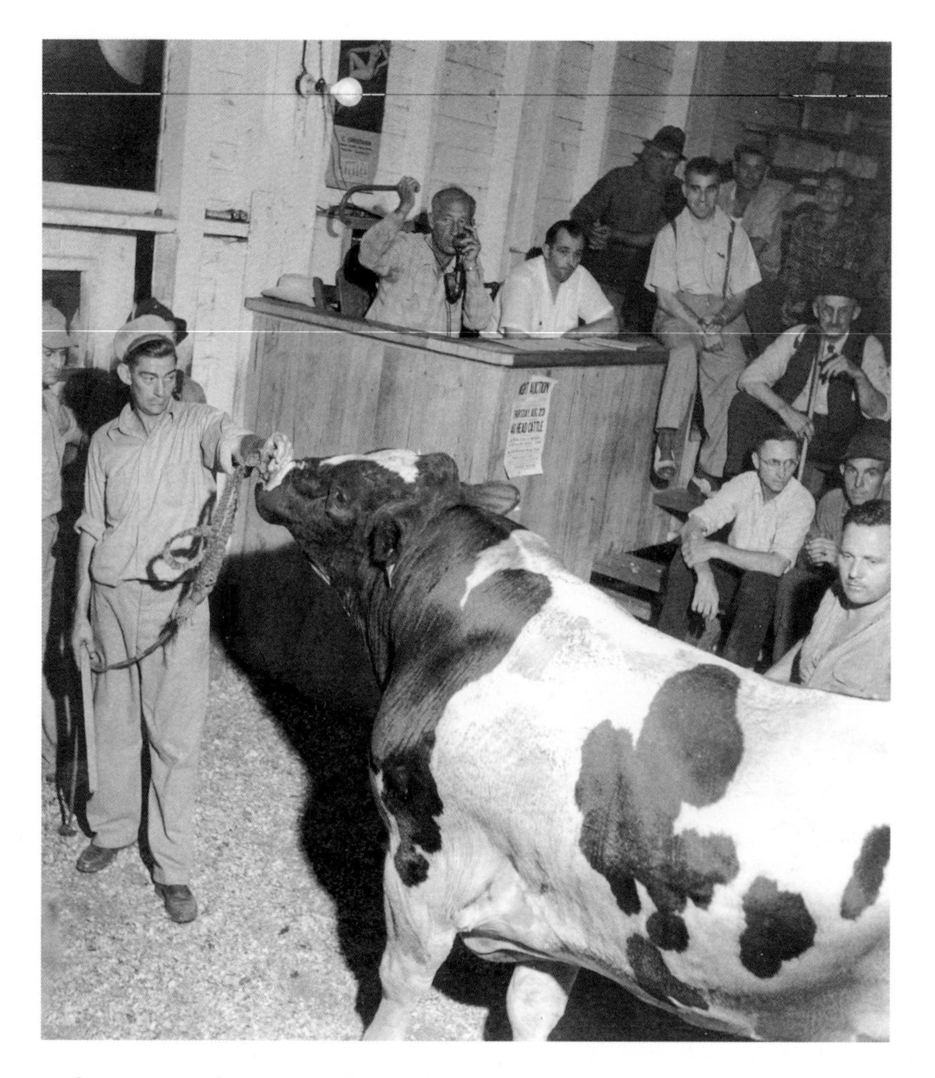

Fig. 106 Holstein bull "put on block," Reinhart's Auction, Canajoharie, New York, August–September 1945. Photograph by Sol Libsohn.

they just go right into squealing and it's the worst noises you ever heard. Their eyes get red. He has completely lost control of himself—that's what happens.

Doug: Now what would bring that on?

Buck: Just a fit of anger. He gets real mad if he sees a stranger, or whatever. Like I picked one up to ship last summer—he was mean; he would chase two brothers—one brother he liked better than the other, but one brother, if he was out in the field and he looked out the door, he would come right out of the barn in a dead run. The other brother could do a little bit with him. But you have to get the trailer all set up to get him.

Doug: How do you get him?

Buck: They have him in the barn when you get there, and you get backed up, and you get your gates with ropes on them in the trailer, so that you can shut the gates from inside. I usually have a time gate on the truck; see I carry the tailgate on the side, that swings open and blocks one side, then you put the farm gate on the other side, and when you turn him loose he has no place to go but through that door and into the trailer. Then you pull the door shut and twist your ropes and stay out of sight until he calms down, because he would tear your trailer apart.

Them big bulls weigh a ton. I loaded three for Glen Sprawells up here—Cold Springs Road. Three weeks ago—yes, three bulls with horns. Glen buys them when they are calves; he throws them outdoors and they never go in the barn again until they weigh 1700—1800 pounds, and they've got horns like this [*Buck spreads his hands*].

Glen wanders around them fields with them bulls—I've been back there with him—and he would say, "Can you help me get that animal out of there?" I should know better. I get back in there and I see these big bulls standing around looking at you, and I say to Glen: "Are they ugly?" "Well they never have been . . ." You know Glen—he's a puffin' and a wheezin' and his teeth are clattering. I am afraid someday he is going to end up dead up there.

Well anyway, we loaded three bulls in the truck three weeks ago and he had a bull and Black Angus. The bull went in with him, so she went in the trailer with the bull, we couldn't get her separated out. So she went in the trailer and all the bulls went in there; well, after five or six minutes I got her situated to where I got her off. Immediately after I got her off, the three bulls got in a fight in the trailer. I mean they were putting the horns into each other and we stuck the cattle prod in there—didn't even fizzle them because they were figuring it was the other bull. I didn't even wait to give Glen his ticket. I said, "I have got to get going or they'll wreck my trailer." Well, they punctured the fender where it rubbed against the tire. You see, you get them on the road and then you can slam the brakes to a certain extent and throw them off balance. Zigzag the truck.

I had to take them right to the sale barn. I had more animals to pick up, but I couldn't do it. They were acting so bad that I couldn't do it. So now when I got his stuff I figure on putting his animals in one compartment, whether it is full or not, because they have horns on them. If I put your cow in there and his bull shoved the horns through her stomach—you're not going to get nothing out of your cow. It ain't fair to you.

So you see, a bull is something you never trust. Especially a little bull. If you're following him up a lane or something, or out in the barnyard bringing him into the barn, if you slip in the mud and fall down, he's just liable to attack you right there. Anytime he catches you off guard or if you go out and grab a calf and the calf lets out a bellow or you go out and rope a heifer and she is fightin' you—anything like that ticks him.

That is the way a bull is. People have the idea that a bull who don't like people will bunt you around, and kill you that way, but a real old bull that knows his business will crush you; he'll drop on his knees. He'll knock you down and drop his knees on you and crush you with the heft of his body. A young bull will punch.

I got tore up pretty good with a bull years ago, and he just rolled me up and down the barn. I'd come to and he would have his head in my chest up against the wall and he'd be on his knees, and then I'd black out, and then I'd come to again. I remember one time I came to and I was looking into that curly spot on his head—he had his head right against me in the wall and I can remember thinking, "If he pushes more I'm done for." Then I blacked out again. We had a big old-fashioned washtub—you have seen these washtubs that they used to wash clothes in—we used it to water heifers. Well, it was laying up against the wall on the barn floor and he rolled me up against that tub, and the tub was completely flat. It caved the sides right in on that tub.

Doug: How did this start?

Buck: He was ugly and he wouldn't come to the barn like a bull; he would make you go out and get him. So I would take a pitchfork or a club—back thirty years ago I thought there wasn't a bull alive I couldn't handle. He would go around and around and circle you, and like you say, if you didn't back off, you could bluff him down, you see. Or you would take a piece of pipe and you would take him across the nose, is the best spot—top of the nose, that was a tender spot. Well, this morning he came in the barn—and I only had a two-foot piece of three-quarter inch pipe—and he'd always come in and pile into the beams. Every four stanchions we had a beam going up. He would come in feeling real obstinate and he would pile into the beams you see. So I took this two-foot piece of pipe and gave him a club across the nose, but it wasn't big enough, so he took after me. I run up through the alley and turned down in front of the cows, and he shot right into the barn, so I shut the door on him.

See, he came in with the cows to take a stanchion. He came right in with the cows. One morning before this, he chased me into the barn floor, so on this side of the barn you had a door that went into the heifer barn; it used to be feed holes for horses.

So I thought, I'll run into this door, spin around, and come right out through the feed hole and shut the door on him. The beams—they were four by fours, probably three foot apart, you see. Two layers of boards. Well I come through the hole—I never got a chance to look back because I could hear the boards breaking—he come right through—he got his head through that hole—he chased me and came right out through the wall. He came through two layers of boards. So I just shut him in there—this time he got me was probably a week later.

We went through this ordeal everyday. So this day I got the cows all in—the hired boy was out there with me—I got the cows tied up and I looked in the barn for him and he was standing there half way down the barn floor chewing his cud. He looked fine to me, you know, so I walks in to go around him, and I got halfway by him and he just charged right into me. I tried to duck him but he caught me in the side and threw me—I think I went as high as the ceiling and when I came down on my back on the cement floor he was right there. I threw my feet out because he was coming at me, and I was on my back, and I threw my feet up and he caught me in the back of the legs. And then I don't remember no more. Billy Shore was with me and he went up in the haymow. He was just a kid like fourteen or fifteen; I was about twenty-five. Billy said he just rolled me up and down the barn; that's how the tub got busted. See, he rolled me from one end of the barn floor to the other, and then I came to a couple times. When he was on top of me I'd come to. . . .

I guess Billy was too scared to get help. My mother was late coming to the barn that morning. She heard me moaning in there. And she knew right off what was going on, so she went to this big old door that hadn't been opened in years, and she yanked enough on it to get it open and looked in there. And when the bull heard the door squeaking—when she looked in, he got up off me and come back, because I wasn't moving, see—he looked out the door. Eventually he would have killed me because when he would've got some blood on him he would've gotten worse. I remember coming to and seeing him looking—he was standing at the other end of the barn floor. I jumped up and got right out of there. Within five minutes, I was so stiff from my neck down to my feet that I was just like a stiff board. I didn't have nothing broke, but I was so stiff—and for a week at night I would go to bed and in the morning I would have to roll out of bed on my knees.

Doug: Nothing broke though?

Buck: No, but I couldn't sit up. If he would've got some blood he would've got worse and that is what old Luther Hargrave told me: "You was lucky that he was a young bull, not an older experienced bull." They used to collect semen from them, you see. And Luther had been taken down I don't know how many times, and he said an older bull would crush you with his knees.

I sold the bull right off, because I couldn't do nothing with him then. I was scared of him and I have been scared ever since. I truck bulls and people say to me, "He is no trouble; you can lead him right on the truck," and I'll say, "Here's the rope; you lead him!" Because I have had bulls that haven't bothered nobody and you run them on that truck, and I don't know if it's the smell of the truck or the smell of the cows or what it is, within two seconds they have gone from a calm tame bull to a raving maniac. So I just don't trust them.

A bull killed Paul Shriver—who lived just down the road—two years ago. He has bought and dealt with cows all his life and he's been around more bulls and cows than I'll probably ever see, and he got killed by a bull. They take chances; they have been around them so long they take chances. They may go up there in the field and drive a hundred bulls in the barn, but it only takes one. It is like driving a motorcycle or driving a car or anything, when you get to the feeling that you have mastered whatever you're doing, that's when you look out.

In Paul's case, they had the bull in the barnyard and he jumped over the fence. The kid in the house or somebody done something to get his attention. Well, the bull went over the fence, so old Paul kind of flew off the handle and he went out to get him. The bull charged him and killed him. In fact, he had an auction two weeks before. He was a dealer; he had the only auction I have ever been to where he had all the hot dogs, coffee, soft drink, hamburgers, donuts all furnished by him at the auction and two weeks later he was killed.

Buck's long story is an account by a farmer who, in 1999, remains committed to less rationalized, less factory-like farming methods. For Buck, farming includes dealing with an animal like a bull; they are part of the natural system of a dairy farm. While nearly everyone applauds the benefits of improved breeding, it does not necessarily follow that all things about the system that existed before artificial insemination were illogical, given certain assumptions. Bulls provided services at relatively low cost, their services were less complicated than artificial insemination, and keeping bulls made it possible to rationalize keeping good cows that were hard to breed.

Culling

A farmer increases milk production two ways. One way is by breeding cows with a higher milk production. The other is to shorten the time the cow is "dry," that is, not producing milk (farmers agree that a cow about to give birth should have some dry time, although it is interesting to note that if a cow is infertile, she may be milked comfortably, without a pregnancy cycle, for as long as four years).[71]

The transitions taking place during the SONJ era were redefining "dry time." A cow becomes dry when the farmer gradually stops milking and feeding a high-protein diet. But if a cow does not become pregnant when first bred, she has a longer period of low or no production at the end of her several-month milking cycle. Each month that the cow is not successfully bred lengthens low- or no-production period. In addition, if a farmer is breeding artificially, costs escalate each month expensive semen does yield a pregnancy. As farmers began to focus more on milk production, they began to treat their cows less as animals and more as units of production. A cow that is hard to breed, in the newly rationalizing system, is a cost not worth bearing. Mason describes the traditional attitude:

Doug: Did you butcher cows that went bad?

Mason: Well, one that you couldn't get bred or something, you know, fat and lazy. But you always got some that won't breed . . .

Doug: How many times would you try to breed her before you would give up on her?

Mason: Oh, you'd turn the bull in with her. Bull would be with her all summer, you know.

Doug: You give that much time?

Mason: Yeah.

Willie: You want 'em to be friends!

Mason: When it come towards spring you could feel calves in 'em, you know. But, the ones you couldn't, that's the ones that went for the beef. You could tell.

The more industrial farms which emerged out of the SONJ era eliminated the bull, the farm dog, and most of the sentiment for the cow. Like all change, it is treated with ambivalence. Treating a cow as a milk machine may make rational sense but it doesn't enrich life or create stories. The old farmers love to talk about bulls because they added a bit of drama to their lives, certainly a degree of uncertainty. They didn't ship a cow for beef the first cycle she didn't become impregnated; they turned her out with the bull for the summer, they "wanted them to be friends"! There is irony in these stories, of course. But the farmers loved to tell them. The industrial farmers I interviewed at the end of my study had concepts to discuss, but few stories to tell.

Herds and 4-H

The gradual specialization of dairy farming around a few breeds was encouraged by the development of 4-H, an organization established in the early twentieth century to promote farming. The motto of 4-H is "learn to do by doing," and their acronym stands for "head, heart, hands, and health." It was a powerful socializing tool in rural areas as well as a mechanism by which young farmers carried on the traditions of farming. Jim Fisher, an active member throughout his life, tells the story:

Jim: 4-H began in 1926 or 1928. Bert Roger was really the founder of 4-H. He was a Cornell graduate and lived over in the Winthrop area, and he established the first 4-H club in the country, and then was hired as an agent, soon after that the whole program got rolling. My parents were leaders, and I was a local leader, an assistant leader, and active for twelve years. It started out with

Fig. 107 Delivering heifer for livestock show, South Lebanon community fair, Lebanon, Pennsylvania, September 1947. Photograph by Sol Libsohn.

livestock and dairy cows for the most part, and then the homemaking and the girls' department came along and now, there are so many different projects! Dairying is one of the many, you might say, because there are so many fewer farms and it's still—I wouldn't say it's a minor one—but it has definitely taken a back seat to all of these other projects.

And so the clubs have changed, you know, the composition of the clubs. They were small to begin with, probably eight to ten members, and there was a lot that you could accomplish, it seemed to me, with a small group of people

. . . well, it's just like small classes and large classes. I know, by the time we gave up being 4-H leaders after about twenty-eight years, Marian took over and soon it was up to fifty members. I attended some of the meetings, and I just said, "Heaven forbid, I'm glad we're out of it, because it's so difficult to work with such large groups!"

But they have organized well, and the leadership is shared by a great many people now. In the beginning, two people, often a husband and wife, were the agricultural and the home-ec leaders and now they have leaders for all the different projects. It's more of a shared thing, which is good with so many more members. But the whole thing is changed from the way we knew it. And their aims and goals are different.

We concentrated on teaching the kids how to grow good animals, and show them. . . . Oh, yes, it's an educational program and supposedly a means of farm fathers working with their kids from a very early age, preparing them for being in the business when they are through.

Doug: Does it still have that function?

Jim: Yes, I would say, it would have that function, because you see most of the good farmers in the country are former 4-H members. I think all of them are . . . not all, but many of them. I can remember at the banquet, Reggie Chester mentioned that his children enjoyed 4-H, but he never was a member, and I was surprised because I thought that everyone in agriculture had been a 4-H member in the last couple of generations. I think you would find that most of the farmers in their midthirties or forties are former 4-H members and have had that background.

It's a great training program. It would bring the fathers and the sons together; it's something they can go and enjoy together. I can remember just so many great times I had with my father going to the state fair. I can remember two different times of him buying a new International three-quarter-ton truck and building a rack just so he could take James to the state fair. So we would go to the fair and some of the other fathers would go along in their trucks—we would have kind of a caravan of trucks just going along together. Some of our best friends are people that were showing and in 4-H with at that time. After the fair, you'd always have the trip home and you'd be planning what you could do for next year—that was the time to plan for next year, right then, and get your plans made and it was a bond between the fathers and sons, and something they took a lot of pride in.

My projects were all dairy, practically, and showing Jersey cattle. So, from starting with one animal that my father had purchased that came from Jersey Island, and raising her progeny, I probably had twenty-five or thirty head by the time I was ready to start farming. It's a way of breaking into the business nicely.

Fig. 108 Stock parade, Bath fair, September 1945. Photograph by Charlotte Brooks.

Doug: You described it as a way of life where the lessons that were essential to what you were going to do for your life's work were passed on to you, and that would seem to be pretty serious business, even though it was not done in a serious way, maybe.

Jim: Yes, I would say that being a good dairy 4-H man is a good indication if you liked it, or if you knew you didn't like it; it was a good way to know.

Emily: Actually, some 4-H members build up quite a sizable stock for the future.

Jim: Yes, often they would have twenty-five or thirty head by the time they were through 4-H. If they have a successful animal and are able to breed and raise some stock and keep all that in their name, by the time they are through high school, or college, they can either sell their stock and go on to college, or they can keep it and have a good nucleus for a herd when they are old enough to begin.

Fig. 109 Prize: Ayrshire bull and owner, Bath Fair, September 1945. Photograph by Charlotte Brooks.

Coincidentally, 4-H recreated the gender worlds of rural women. Emily Fisher's comments were inspired by Brooks's photo, figure 93.

Emily: I oversaw mainly cooking and homemaking projects. I didn't do too much with sewing—I had little interest in sewing. But I liked to do cooking and house jobs. The 4-H has a curriculum you go through. It's a set program. They start out with simple sewing and cooking, and then progress through the different years. They have hospitality and correct manners and that type of thing built into all of the courses. Kind of an overall preparation for being a good homemaker.

Jim: This is a photo of the fair (fig. 107); these cattle look as if they're well groomed, and this yes looks like a fair scene. They're unloading.

State fair usually comes along starting Labor Day; runs a week after most of the county fairs are over because you participate in the county fairs, and then the winners from that are chosen to go to the state fair. Those are open class in the junior divisions.

Doug: This was a big part of your year when you were a youngster?

Jim: Oh yes. I started showing in 1925 and have exhibited every year since. This will be the first year I haven't—well, we might have one lamb there this year.

Doug: Would you begin to build a herd as a young person through these projects?

Jim: Yes. We started our whole Jersey enterprise through my 4-H project. We had Ayrshire cattle for a number of years, and we went through a stage were we had Guernsey cattle, but disease in those years often wiped out whole herds of cattle. We lost all of our herd with tuberculosis one time when they started testing for TB. All the cattle that reacted to TB had to be slaughtered and sold. That would have been in the twenties. Everyone in the area was affected—there was an horrendous outbreak of tuberculosis. Then that sort of cleared up. It was at the time when people were infected with tuberculosis, and often, if you had a worker who had tuberculosis and he was spitting and feeding the cows—that was the way the disease got spread. Then after that was cleared up, the next epidemic was brucellosis. We had bought a whole group of cows and didn't realize that they were infected, and they began losing their calves about four or five months along. They were tested, and they had to go, so that was another big loss for dairymen.

Doug: These wouldn't just wipe the farm out?

Jim: No, you would bounce back, because there wasn't much money invested in them at that time—but it was a bad shock. We went to Ohio and brought back Jersey replacements in the late twenties, and then my father bought me a Jersey yearling that was imported from Jersey Island. She was a beautiful cow and she showed well. She was first at the state fair, and she went to the national Jersey show in St. Louis and was first prize senior yearling and reserve champion. That was 1928. A 4-H team or a group of cattle went from NYS—one of my best friends from Gouverneur, Herb Putnam, had an Ayrshire cow that went to the same show. The cattle went on freight cars, and the young people followed in the train.

So that was an interesting experience for kids, and through that one cow we gradually expanded. We saved one of her early sons as a herd sire and then added to that, and built of a herd of prize Jerseys. He was a sire for quite a bit of the herd, until he was related to so many we had to do outcrosses. That was before artificial breeding. Bull associations were formed, and farmers switched

> bulls—and they were able to prolong the life and save buying expensive herd sires. So you formed a bull co-op and traded bulls with your neighbors. The only problem with that was that, with a big heavy sire, it was difficult to breed a yearling, because they were so heavy that the yearling wouldn't hold the bull. But the extension service helped the farmers build a special bull yard and a breeding stall. This was so that a young animal could be put into the breeding stall with a large animal. There was a ramp on either side so that when a large animal mounted the small cow his front feet came down on a ramp and his whole weight didn't land on the animal. In that way you were able to use a heavy old sire on a young animal, and you extended the life of the bull by trading him around—we did that with four or five neighbors that had Jersey cattle.

These stories of feeding practices and decisions, and the role of 4-H in socializing new generations, are interrelated. In the pre-factory-farm system, the force of tradition led to conservative decisions. Work was designed and paced by humans. The pace varied; it slowed down in the winter and assumed an almost relaxing rhythm when the weather was nastiest. In the summer, farmers worked most light hours and some in the dark. Much of this work was shared with neighbors, and that lightened the load. Because the scale of farming allowed farmers to know their animals individually, herds were built with subtle characteristics.

Housing and Milking Cows

There is a final topic in this portrait of change, and that concerns barns, milking systems, and the use of manure. Farmers moved animals indoors in the early nineteenth century and designed barns that are still in use on many farms. These barns are two-story buildings with a hip roof. The top floor is used for hay storage. The first floor of the barn is organized so that two rows of cows stand with their heads in stanchions, facing the outside of the barn. A farmer feeds the cow from an aisle in the front of the cows and cleans manure that falls into a gutter behind the cows.

Animals in these housing situations spend much of their lives outside and proceed into the barn and to their customary stanchions where they are milked and fed. In the winter, they spend most of the time inside. Farmers spread straw on the floor for bedding; as the straw is soiled it is removed, creating a manure with fiber and bulk. The stanchion barn is actually a rather complex space, with various rooms for feed storage, calf housing, and, in the beginning, tie stalls for horses.

For a person who has never been on a dairy farm, the stanchion barn might be a surprising place. The animals stand calmly in line, munching grain or hay, occasionally mooing. During milking there is a quiet background sound of the milking machine, pumping quietly. The stanchion barn is orderly. The work consists of milking cows, shoveling feed or manure, and studying each cow for indications of health problems or monthly fertility. Of course, there are well-ordered and clean stanchion barns and those that are terribly disorganized and filthy. The quality they all share is that the structure of the building assigns each animal a space, and that work in these barns means moving from one animal to another. In this way the stanchion barn is like the factories that preceded the assembly line; the worker controls the pace of work and asserts his or her mastery over each act of production.

The stanchion barns of many contemporary farmers is essentially similar to those photographed during the postwar era. I will explore the implications of transitions from these barns to free-stalls following sections.

Milking Systems

The original milking machine was actually a milkmaid who milked a small herd, which was wandering in the barnyard. Women were thought to be more patient with cows, because they understood the procreative experience. There were typically only a handful of cows to milk. The milkmaid is a common figure in nineteenth-century American and European novels describing preindustrial rural life (Thomas Hardy's *Tess of the D'Urbervilles* is a fine example). When milking cows became an important part of the farm, the work was desexualized, and, for the most part, men assumed responsibility for it. As milking cows has been done on a larger and larger scale, the processes have become more like mass production and less like a craft activity.

Milking a cow is essentially a rhythmic pulling and squeezing of the cows teats, shooting out a stream of milk until each section of the udder is depleted. When I took my rural sociology students to a North Country farm, my neighbor had each student try to milk a particularly patient cow. He said to the women in the class, "just pretend it's your boyfriend!" Given the typical rural mentality, he did not tell the boys, "just pretend it's yourself!" Indeed, the students got the hang of it quite quickly. I have shaken hands with many elderly farmers who milked ten to twenty cows a day for decades, and their hands are extraordinarily strong, but often bent into half-formed fists. Most people do not remember hand milking fondly and were glad when the first mechanical milkers came into use during the 1920s. These machines fit over the cow's teats and sucked the milk either into a pail or through pipelines to milk cans, or, eventually, to a bulk tank. It remained necessary for farmers to milk by hand occasionally, when a cow was too tender for the machine, or too stubborn.

During the SONJ era, farmers stored their milk in milk cans, which they delivered to local plants. As is evident in these images (figs. 112–15), this work consisted of lifting, pouring, and jostling heavy milk cans. Handling the milk in this manner

Fig. 110 James B. Stewart plugging in electric milker, Brooklea Farm, Kanoa, New York, September 1945. Photograph by Charlotte Brooks.

exposed it to potential contaminants. Probably the most positive aspect of this system was that milk delivery was a daily ritual, which often brought farmers into cooperative arrangements. Some of that energy is evident in these photographs. In fact, it looks like it is fun to get the Model T out for a daily drive to deliver cans of milk on a beautiful summer day. A daily drive took the farmer across the borders of his farm and offered the chance to pass easy time with his neighbors. I am sure it was less enjoyable during the cold weather, but, as we noted previously, the production of milk dropped in the winter to little or nothing.

The milk can system depended on local milk-processing plants. These local plants made fluid milk (here the processes were pasteurization and packaging), butter, or cheese. Local cheese plants produced food with distinctive characteristics, which are still evident in dairy areas in Europe but are now rare in the United States. Modern cheese making is based on national markets and produces a bland and uniform food.

Fig. 111 Andrew Shuman, Jr., milking, Shuman Bros. Farm, Roxbury, New York, September 1945. Individual cows give an average 6,000 pounds a year, while some give as much as 12,000. Photograph by Sol Libsohn.

Just after the SONJ era, canned milk systems were replaced by "bulk tanks"—refrigerated stainless steel containers ranging in size from 500 to several thousand gallons. The farmer dumps milk into these tanks until a truck from the milk company comes to the farm to pump it dry. It is a step in the process of rationalization and industrialization which eliminated the limit on the size of the milking herd, imposed by the practicalities of moving milk cans. Indeed, with the bulk tank system there is no practical limit on the number of cows one may milk; the farmer needs only to purchase a larger bulk tank as his herd increases, or arrange for more

Fig. 112 Milking time on Lang's Gold Brook Farm, Stowe, Vermont, March 1947. Farmhand is straining the milk in the milkhouse. Photograph by Arnold Eagle.

frequent visits from the company milk truck. The bulk tank also eliminated the work of moving milk by hand in tanks.

The refrigerated bulk tank was expensive. It only made sense to have a bulk tank for a herd of a certain size; therefore, bulk tanks meant that farmers had to either increase their herds or cease farming. Coincidentally, the trucks themselves, which were driven from farm to farm to pick up milk, had an impact on who could stay in business. Many small farms in the North Country were accessible only by small bridges that crossed creeks and small rivers. These bridges were often not strong or wide enough for the milk trucks; those farms on the other sides of these bridges were left out of the bulk tank system. Most sold their herds.

Finally, the bulk tanks made fluid milk easier to move longer distances. It was already in a large truck; shipping it to regional centers for processing and further distribution was a natural extension of this system. The milk truck pictured in figure 116 is the beginning of the rapid change from a local to a regional and national dairy market. Supplying the needs of growing cities is clearly the structural frame for these changes.

Fig. 113 Farmer, New York State, March 1946. Photograph by John Vachon.

Fig. 114 Farm truck on Route 5 South, Fort Plain, New York, August 1945. Photograph by Sol Libsohn.

The farmers I interviewed experienced this transition in several ways. Virgil, a small farmer who used a stanchion barn his whole life as a farmer, shared a bulk tank with a neighbor:

> My neighbor down here was the first one in the area to get a bulk tank. You carried the milk in and dumped it. For awhile, I didn't have enough milk for the pickup and my neighbor brought his milk down and dumped it into mine. They got picked up together. I kept my quota up; they won't come and pick it up if you don't have enough. They'd come every other day.

The first bulk tanks were financed by the milk companies. Virgil explains:

> When we first got the bulk tank, all we had to pay was the freight on it to get it here from out in Wisconsin someplace. All we had to pay was the freight on it. And then they took out ten cents a hundred—I think the price of it was five hundred, and they hauled the milk for three years free . . . that's why we got in on it . . . and the neighbor down below got in on it.

Jim Fisher describes how the transition to bulk tanks influenced the local milk economy:

> *Doug:* Was there a sudden change then in the way milk was collected? Was it suddenly impossible to distribute milk in milk cans?

Fig. 115 Farmer getting back his empty milk cans after delivery at Horseheads, New York, receiving station, March 1945. Photograph by John Vachon.

Fig. 116 Truckload of milk brought to Middletown from further north will change drivers before it is shipped into the city, March 1945. Photograph by John Vachon.

Jim: The local plants went out of business with the bulk tanks. Once the bulk tanks started, the milk companies found that it was more efficient to have fewer plants, and to truck the milk farther. So they more or less called the shots. Eventually there was not going to be canned milk. The farmers would have to put it into bulk tanks, which really eliminated the farmer with fifteen–twenty cows. It came rather slowly, for a while.

Doug: How could those two systems coincide—if bulk tank collection had begun and you had only one or fifteen cows, what would you do with your milk?

Jim: There are a few cheese plants operating, and now the Amish have opened plants if you want to farm on a small scale again. So, actually, there is more opportunity again, now, because of the Amish, than there was for a little period of time. I think that there was always a cheese plant in Hermon or Russell, or one of those areas.

Actually, I was pleased with the bulk tank. I had gotten to the point where I hated hauling milk cans in and out of the cooler. That was—for me—a good change because it eliminated a lot of heavy work. The milk was cooled immediately and the milk was of much better quality, I felt, than the milk that we shipped from the farm that was cooled slowly in nothing but a cold water bath around the milk cans. I had no gripe on the bulk tanks themselves.

The bulk tank, like many technologies, brought complex changes to farming. All farmers applauded the labor saved and improved milk hygiene brought by the bulk tanks. Yet, like many technologies, it led to perhaps unanticipated consequences. These include the exclusion of small farms from a system which has a market only for bulk-tank-stored milk. Buying a bulk tank required that farmers expand their herds, and it made further expansion possible. And while hauling milk before the era of the bulk tanks was hard work, it was a job that farmers often did together, and a job which got them off the farm on a daily basis. Said another way, greater productivity narrowed the tasks of production and isolated farmers.

I end my discussion of the SONJ era with two summaries. The first examines records that a farmer loaned me for my research several years ago. These were meticulous handwritten records of all income and expenses from 1939 to 1941. I summarized aspects of these extraordinary eighty-five pages of information (the milk chart shown as fig. 97 is from these records) in table 3.

The farm, milking twenty-two cows, was making money. In 1939 the farm netted $2,852.70; by 1941 the income had increased to $4,150. The price of milk during that era, just on the brink of the U.S. involvement in World War II, had increased from $1.65/hundredweight to $2.28. One imagines that this rapid inflation had to do with the oncoming war.

The average milk per cow at 8,600 pounds/year is about a third lower than that produced by current cows. There are several ways of looking at this. Part of the increase in production is due to the shift from mixed herds which include Jerseys (which make less milk but of higher quality) to the pure Holstein herds of the modern farmer. Most of the increase in production in the past fifty years has to do with successful breeding and improved feeds.

The breeding activity of the early farm is reflected in the farmer's sale of nineteen cows and calves during the year, and the purchase of four cows. A bull calf brought $40; female calves brought from $5 to $11. Three bulls averaged slightly over $100 each. Cows sold for about $100, except for a Jersey, which fetched $275, and a registered cow, which brought $140. The farmer sold nineteen cows and calves for $1,201 and purchased four cows at a cost of $406.

The farm of the 1940s produced for both self-sufficiency and income. One hundred seventy-five chickens provided much of the food for the family as well as income from egg sales. In fact, the egg sales paid for the chickens the family consumed and made a profit. The family also raised turkeys, pigs, and lambs and sold barley as a cash crop. These supplemental animals and crops provided more than 10 percent of the farm income. Garden produce and other crops such as honey, maple syrup, blueberries, and raspberries are mentioned in several ways in the records but not listed as income. They undoubtedly enriched the family diet.

TABLE 3 Income and Expenses for a North Country Dairy Farm in 1939

	Expenses($)	Income($)	
Cow feed (22 milkers)	589.00	189,864 lbs. milk (8,630 avg./cow)	2,882.36
Cows purchased (4)	406.00	Calves, veal calves, cows, bulls sold	797.94
Hauling milk	57.71		
Pasturing heifers	28.00		
Sheep purchased	80.55	Lambs sold	90.50
Chickens purchased (175 + feed)	126.08	Eggs and chickens sold	290.54
Turkeys purchased (64 eggs + feed)	77.77	Turkeys sold	157.04
Hog feed	11.20	Pork sold	32.68
Feed grinding (oats and wheat)	66.90	Barley sold (7 bushels)	8.25
Ice and sawdust for ice	8.04	Wood sold	75.50
Seed potatoes	13.64	Soil conservation	61.65
Barley seed (4 bushels)	5.00		
Limestone, phosphate	9.00		
Veterinarian (2 visits)	9.50		
Labor (avg. $1/day)	166.70		
1/4 beef	14.75		
Total	1,669.84		4,396.46
Net			2,726.62

Source: private farm records.

In 1939, threshing cost $10.50; by 1941 the cost has inexplicably increased to $27. As Arthur described, farmers paid workers about a dollar a day; for each of the three years of the records I analyzed, the farm paid out about $100 to seven to ten different workers. The farmer paid $36 for the services of the milk tester, pictured in figure 103. Fertilizer cost $113. The farmer paid $600 for cow feed, which included mixed feed for dry stock, and the cost of grinding oats.

They bought no equipment during the three years, except a $30 wheelbarrow and a $45 spring tooth. There are truck expenses listed, but no tractor expenses. It is likely they had yet to purchase their first tractor.

It was an advanced farm for its day, using DHIA testing and raising at least some registered cattle. But the 1939 farm is remarkably similar to a farm a hundred years before, and it resembles the Amish farms of today. The key elements are diversification, family self-sufficiency, a profitable milk income, and overall profitability.

But in three years the farm changed a great deal. By 1941, farm net income had increased by a third. A great deal of the increased income was due to milk increasing from $165/hundredweight in 1939 to $2.28 in 1941, undoubtedly because of economic uncertainties related to the U.S. entry into the war. But the farm had suddenly become less diverse, eliminating poultry and other secondary crops. The future had begun.

Fig. 117 Aerial view: farm near Hoosick Falls, upstate New York, 1949. Photograph by Charles Rotkin.

I add to these data Rotkin's visual summary: a 1949 aerial view of a New York farmstead.

The photograph shows several small outbuildings, which were probably used for pigs, chickens, and perhaps turkeys. The silo held corn from a ten-to-fifteen-acre field, most likely harvested by neighbors. Hay for twenty to thirty cows the farmer probably milked is stored in the mow of the barn.

The other buildings spin a bit of mystery. The building in the middle of the farmstead could be a new equipment shed; it could be a horse barn or blacksmith's shop, or perhaps a large well house with cement cooling tanks for milk cans waiting to be taken to market. Surely the small building just along side is the doghouse; cows still went to pasture, and the dog helped get them there and back.

The house is comfortably away from the barn, shaded by elms or maples. The farm was probably established in the nineteenth century and it probably now holds an elderly generation who spent their whole lives there and perhaps also a generation fresh from World War II and the social change it promised. This is the world I have described in the preceding pages, with the help of the SONJ photographs and the farmers' tales.

The History since Then

Fig. 118 *(previous pages)* Full-moon milking on a factory farm, –25 F, 1990. On North Country megadairies cows are milked three times a day. Milking is shift work which goes around the clock. Photograph by the author.

In 1989–90, I interviewed forty-eight of my neighboring farmers with the intent of understanding farming from the perspective of a neighborhood system.[1] It was a sabbatical year, a wonderful year of largely uninterrupted fieldwork, and I can still smell the manure on my boots.

I identified the farmers by obtaining a list from the local cooperative extension of all farms shipping milk in a section of the county with which I was familiar. I used a plot map of the area to locate fifty contiguous farms. I wrote each of these farmers a letter identifying myself as a sociology professor interested in understanding the "problems facing local farmers" and asked for their help in completing a survey I would bring to their farms. The letters went out; follow-up calls were made, and most of them gruffly agreed to see me. I am not comfortable asking strangers for something that will bring them little in return, and I scheduled my interviews so that I could offer each farmer some modest labor. I asked if I might be present during a milking, realizing that a stranger sometimes spooks cows, which causes them to constrict their milk supply. Most farmers invited me to the barn anyway.

On many of the farms, there were tasks I could do. I threw bales of hay from the mows and fed cows or calves. Or I shoveled manure from the calf pens, since it was a leftover job which nobody really liked. I didn't mind shoveling a little manure, but I needed to learn the etiquette and health considerations of when and where to clean my barn boots. There were other jobs I did, but I never milked the cows, and I did not feed them grain or concentrates; these tasks were too important to assign to a visitor.

I found that most farmers enjoyed a visitor who was willing to pitch in. I liked being in the barns and I learned a great deal from the casual conversations I had with farmers as they did their daily chores. I usually arrived for the morning milking (usually between 5:00 and 7:00 A.M.) and was invited to the house to have breakfast after the milking. By the time we had finished breakfast, we were usually ready for the survey. Sometimes I went with farmers on errands during the day, as the survey took place during the winter and there was no work in the fields. In some cases, I stayed through the chores such as manure spreading, the afternoon milking, and even into the evening. When I returned from these experiences, I would immediately write out my observations. Reading these field notes as I prepare this manuscript transports me to those good barns and wonderful people.

I did not have these kinds of experiences with all the farmers I interviewed, of course. Some farmers treated the survey in an entirely businesslike fashion and did not invite me to the barns. In these cases, I was in and out in a couple of hours. On the larger farms there were no informal tasks I could do because there were hired workers assigned to the jobs. I was more of a sore thumb and decidedly less of a *participant*/observer.

But in all these were rich experiences—the great moments of fieldwork. My final ideas about farming draw on the 1989–90 survey and subsequent interviews, and a follow-up survey with key informants in 1999, in which I determined which farmers remained as farmers, who had expanded, and why those who had left farming had done so. Looking back ten years, we see the 1989 data as a point between the SONJ era and the present.

The most obvious message from table 4 is that I came to see the farm neighborhood as consisting of two types of farms, which I refer to as "craft-like" and "factory-like." The craft farms have a direct lineage to the SONJ farms; the factory farms are organized similarly to the truly industrialized farms of the American South and appear to be the harbinger of the future of dairy farming in the northern regions of the United States.[2] I note that these are "ideal types" in the sense meant by Max Weber and best characterize the extremes of the two systems.[3]

My sense that there were two types of farms emerged gradually.[4] The differences I noted were both structural and cultural, including technologies (for example, milking and manure systems and barn types), the consideration of gender and other factors in the organization of labor, and attitudes expressed informally and in response to my questionnaire. Some farms were hard to classify (for example, one small farm, the size of a craft farm, used a free-stall and some "industrial" techniques; see fig. 130), but most farms fell rather easily into one of my types, as reflected in table 4.

To place these data in context, I update some of the history to show how many technological changes had affected all farms since the SONJ era. Among the most important was the bulk milk storage tank, which, as previously described, allows farmers to increase their milk production without increasing the work of transporting milk.

TABLE 4 Characteristics of Selected Dairy Farms, 1989

	Craft Farms (n = 34)	Factory Farms (n = 12)
No. of cows	14–106	55–268
Mean no. of cows milking	46	108
Annual calf mortality (%)	3.7	6
No. of farms milking three times per day (%)	6	38
Tillable land in production (acres)	205	412
Corn (acres)	54	146
Pasture/cow	.82	.46
Liquid manure storage (%)	11	54
No. of full-time hired laborers	.37 (1 per 125 cows)	1.7 (1 per 63 cows)
Total equipment horsepower	210	454

Source: St. Lawrence County survey.

The bulk tank was adopted by all farmers in my study and is used by nearly all modern dairy farmers. As one might expect, however, the bulk tanks of the largest farms were several times the size of the tanks on the small craft farms, and it was typical that smaller farms purchased the used milk tanks as their neighbors built bigger systems.

Similarly, as described earlier, all farmers adopted field choppers at the end of the SONJ era, which ended changing works and made each farm capable of harvesting its crops alone. Since the 1950s choppers have become larger and more powerful, and with an easier harvest brought by the chopper, most farmers increased corn production.

The craft farmers, however, use updated versions of the same machines introduced during the SONJ era (often these *are* the same machines, maintained by mechanics like my research assistant and informant Willie), while most of the factory farmers have purchased self-propelled choppers that are larger than the biggest tractor in the neighborhood and cost about $250,000.

Cropping practices between the craft and factory farms are generally similar in that all farmers grow mostly grass and corn, although the factory farmers grow corn on a larger percentage of their tillable fields, and they have nearly eliminated pasture, considering it an inefficient use of land.

Despite these variations, many farmers in 1989 were farming essentially as they had in the late 1950s: they were making hay bales (now often with round balers, which make larger bales much faster); growing corn, which they chopped for silage (silage is now often stored in bunkers covered with plastic); feeding cows housed in stanchion barns; spreading manure with machinery developed around World War II; and shipping milk for one of many uses. Social contact between farmers had declined as formal labor exchanges disappeared, and, consequently, there remained few reasons to share resources, ideas, or fellowship. But despite change that affected all farms, I came to see that within this evolution there were two distinct paths.

Craft and Factory Farms, circa 1989

On what I am calling the craft farm, the farmers make milk something like a craftsman makes a piece of furniture: an individual performs all tasks in the production system, and thus his or her knowledge of the overall system is comprehensive. The farmer uses more rudimentary tools and machines, and the worker controls and directs the machines rather than vice versa.

On the factory farm, workers make milk in a rationalized process which employs an advanced division of labor and mass-production technology. This produces work which resembles factory work, more directed by and controlled by machines. I do note that these differences among craft and factory farmers are less dramatic than between a craft and factory worker making a product. Still, I argue that the metaphor explains a great deal of the character of the current system.

Table 4 reveals some of the immediate differences between the two types: craft farms are almost all smaller and milk fewer cows, they keep cows longer and use more pasture and fewer workers per cow, twice as many of their calves survive, and they have smaller and less powerful farm machinery. These are not simply smaller farms, however; technologies, strategies, and perspectives on farming are decidedly different.

Feeding and Cropping

The craft farmer mixes each cow's feed on the spot, the way a cook prepares a soup: a bit of several grains, vitamin supplements, even ingredients such as molasses to suit the needs and tastes of individual animals. While both craft and factory farmers buy some of their feed, the craft farmers are less likely to purchase high-energy supplements and less likely to "ration balance" their cow's feed (in ration balancing, experts measure the quality of feed grown on the farm and sell protein and other additives to compensate for deficiencies). The result is lower milk production per cow (less efficiency) and less stress on the animal.

A farmer who fits the craft farm profile explained:

> *Tom:* I think whether it's in a bigger operation or smaller one, the closer you push her to her maximum potential the more likely she is to have problems. It's going to put more stresses on her.
>
> *Doug:* How do you push a cow harder?
>
> *Tom:* The better the quality of the feed you give her the more she's going to want to eat. And the more she eats the more she's going to produce. See, the

Figs. 119–20 Two farms organized on the craft logic. Fig. 119: The farm supports the families of two brothers milking eighty-five Jerseys. The farm uses the original barns and several of the outbuildings. In front yard of the original farmhouse is a swimming pond. Photograph by the author. Fig. 120: At the time of the photograph, the farm supported a family milking sixty Holsteins, which averaged 15,300 pounds per year. The herd average was slightly below the state average, as a result of not immediately shipping cows that did not become pregnant on the first cycle they were impregnated (each unsuccessful pregnancy resulting in twenty-one days lost in the milking cycle) and feeding strategies that did not overstress the cows. The original buildings were well maintained and the barn interior was orderly. Photograph by the author.

maximum amount of milk you can make is when all the nutrients are in the proper amounts. So the limitations to any cow's milk production might be any one of these nutrients. When we play the game of balancing the rations, we're trying to bring 'em all up to the proper levels. And so, the closer you get these levels to the optimum amount, the more you're . . . you'd say "pushing" the cow.

Doug: Like an engine that is built for high compression?

Tom: Right. And the fuel mixture. And the air mixture proper. Then you're going to reach your maximum potential. Over the years we've been so far below potential of the cows that you could increase certain nutrients and others could still be lagging and the production would still increase. I think this is where you get your "burnout," where you use the cows up. You increase a certain nutrient which pushes her hard, that way encouraging her to make production, but there may be some minor nutrient that is lacking. It can create foot problems with some kinds of things, or breeding problems with other things. It creates imbalances. And because a cow is working hard—in other words, her stomach is digesting all the time rather than 75 percent of the time. Anything that disrupts her is going to have a greater affect on her body and her functions than previously. You see, if the cow isn't working that hard she can get a disruption and maybe withstand it and not even know it.

It's like if you get everything perfectly balanced, and get this cow producing at 95 percent efficiency, most likely she won't last as long as one who is working at a slower rate. It's like if you take an engine and run it full speed, you know the whole time you have it it's not going to last as long as if you're moderate with it. There's no time for recuperation and so forth.

The most factory-like farms, on the other hand, ration balanced their feed to achieve the most efficient food for a herd. This increases efficiency as well as costs. In the most advanced feeding systems, cows are implanted with a computer chip that triggers a feeding mechanism that delivers optimal feed to a cow. Like many computer applications, efficiency is gained at the cost the human judgment. While the formula addresses the needs of most cows, it may do less well on cows whose characteristics fall beyond the normal ranges. Finally, the computer system is another level of technology with uncertain reliability. Carmen had the first one in the neighborhood but had it removed after a few troublesome years.

Two of the forty-six farmers from my study, both of which I described as working in the "craft" mode, continued to raise grains as an additive for cow feed. One farmer had made a combine from two used machines he had bought at a farm auction and used it to harvest his grains, as had been done by most farmers until the late 1950s. He planted grain as well as legumes to rotate fields out of corn production, both to rest his soil from corn production and to replenish its nutrients without inorganic fertilizer. He sought to lower his feed costs by growing a higher percentage of his cow feed, accepting lower production from his cows due to his less-than-perfect ration balancing. He was, indeed, a modern farmer looking to the past for answers. Other farmers spoke admiringly of his experiments, but he was alone in his methods.

The cropping practices of the craft farmers were less damaging to the land. As pic-

Fig. 121 Stanchion barn, 1993. Notice that the cows are smiling. Photograph by the author.

tured, many of the North Country fields are small and tucked into woods and hills. Small tractors can treat these spaces with more care, because of their maneuverability and relative light weight. Because factory farmers have to get a great deal of field work done in the brief windows of meteorological opportunity, their larger equipment has to be run faster and is less attuned to the needs of specific fields. They might have used contour plowing in the past, or preserved tree lines to prevent wind erosion. The need to get the increased field work done in as few days as possible eliminated these time-consuming but more environmentally friendly methods.[5]

Barns and Manure

It was in the barns and the treatment of the animals that I saw a great difference between the craft and factory farms, and a profound shift from the traditions represented in the SONJ-era photographs.

The craft farmers continued to use the stanchion barns, pictured in the SONJ photos in figure 102 and in my photos in figure 121. Certainly these barns varied in quality, but the best supported work that seemed almost monastic. The cows stand at their habitual stanchions; the farmer moves from animal to animal, attaching milking machines. In the background are sounds of mooing, and the regular pulsing of air through a vacuum system that pumps the milk from each animal to a bulk tank. The smell is a pungent and not unpleasant combination of fermented feed (which resembles a heavy brew such as stout), several other feeds, and manure. Milking proceeds at a pace set by the natural capacity of the animal. You can't speed it up. While the animal is being milked, the farmer observes several aspects of her health and often evaluates how breeding decisions that balanced feet, udders, production, and personality worked out this time around. The farmer greets an animal he or she will know for as many as ten years, through at least two contacts a day, for more than three hundred days a year. All these aspects of work on the craft farm are due, most directly, to the structure of the barn.

There were also stanchion barns that begged for a barn fire (with animals outside, of course!). These were original barns made for smaller cows, perhaps Jerseys, and into which Holsteins, one-third larger than Jerseys, were squeezed. Many had cracked beams from overloaded mows, antiquated manure unloaders, and inconvenient layouts. There was often inadequate housing for heifers or calves. In these barns, which were being used long past their day, one could feel the system in great strain. But farmers still consider the stanchion barn viable and desirable; several new stanchion barns have been built during the time of my study.

A farmer explained some of the ways in which barn type influences animal care:

> I think that's one big advantage to the stanchion barn as opposed to free-stall housing. You're in contact with the individual cows. And in close contact more. The bigger you get, the harder it is to have that contact. There are individuals who get good herd averages because they pay attention to detail . . . they can observe a hundred cows and see which ones are operating normally or not. But here it's much easier to pick up when a cow starts to lose her appetite. That's the first sign of something going wrong, if she's not eating right. And when you're feeding everyday in the same spot and the cow has a certain area that she can clean up, you can notice changes from one time to the other. There are ways of doing it in a free-stall barn, but it takes a lot more observation and a lot more skilled eye.

I spent a morning with another farmer who seemed to be the quintessential cow man. He lived in a small universe that he knew incredibly well (he mentioned having missed five or six milkings in twenty-some years—a terrifying or gratifying thought, depending on perspective!). He did a lot of small things to achieve his slightly above-par herd average. For example, he sheared his cows' hair in winter to keep them cool. Few farmers did that. He pointed to a cow and said:

> Last week her sides were heaving; you would think she had just been chased

Figs. 122–27 Spreading solid manure, 1990. The manure had been gathered in the barn and dumped into the spreader. We drive from the barn (122) and begin spreading (123, 124). These photos are taken from the tractor looking backward, standing unsteadily on the hitch as we banged along the field. The farmer was certain to clean the spreader so that manure would not freeze in the machine, causing it to break when it was turned on. It is a hard job on a cold day.

The final photo in this sequence is an aerial view of spreading manure in the winter, taken on another occasion. The dark bands, as one might guess, are recently spread manure. Photographs by the author.

> by a dog. It is the worst thing to have a cow too hot. She'll lose ten pounds [of milk production] easy per day. When I bring in the heifers I shear all of them; the milk goes right up.

His attentiveness to each animal was apparent, as he milked his cows for the several thousandth time. He treated his animals with interest and engagement; each was an individual project. One cow had stepped on one of her teats; he milked that quarter of the udder by hand and noted that it was recovering. He had had to lance the udder, but was pleased that the cow had thus far avoided mastitis. Another was eating strangely, and he made a note to check what might be causing a change in her behavior. Indeed, even the moods of the animals were noted by the farmer.

The craft farmers had knowledge of their cows which led to individual breeding decisions, rather than using computer-rendered analysis of objective criteria such as milk production and quality to automatically select semen from the AI (artificial insemination) co-operative. The following comments from my field notes reflect this perspective:

> As I enter the barnyard, I see about eighty Holsteins, back from the pasture and waiting to be milked. They are milling about and look impatient. To me they comprise a sea of black and white: big animals all about the same in appearance and mood. But the farmer Charlie sees a cow humping another and says to the eldest son: "Angie's coming into heat; call AI [the artificial insemination service] tonight." The son replied that she was three days early. I was startled by the father's recognition of Angie, and the son's precise knowledge of her fertility cycle. Inspired by Angie, the father and sons then had spirited conversations about the lineage and inherited characteristics of several cows and their offspring. Whether a particular sperm for Angie would produce a cow with greater milk production but a nasty personality consumed several minutes of conversation.
>
> They had named all the cows. They had a hard time finding more names as the herd grew and evolved. A daughter said she had named a black cow Whitey because she damn well felt like it. Jim, the second son, was taking Spanish at the local college and was naming cows Blanco, Negro, Marguerita, and so forth. It seemed a little odd in the blustery November afternoon, but rather more intimate than the numbers tattooed on the cows on the factory farms.

This connection between animal and farmer expressed itself in some farmers' attitudes toward the rewards they should expect for their work. One farmer, like many I would refer to as "organic intellectuals," said there should be a larger bargain be-

Figs. 128–29 The small, peculiar fields of the North Country, 1988. Fig 128: A three-to-five-acre field tucked into woods. Fig 129: One of the few remaining elm trees after the devastation of Dutch elm disease in the North Country. Inexplicably, the tree remains in the center of a large field. It is the kind of local ornament that has symbolic value, but is difficult to work around. Photographs by the author.

tween the farmer and society. He was willing to experiment to make healthy food in an environmentally friendly manner, and at the lowest cost, but he felt his product should be assigned a "just price," just as products and producers had been related in the guild system of medieval Europe.

Another craft farmer, who had sold his herd in a dairy buyout (a controversial program in which dairy herds were slaughtered for meat because of temporary overproduction of milk), described how he felt about what he had done:

> The neighbors felt poorly of us for selling the herd in the buyout. But not so badly as I felt about it myself. I didn't so much mind seeing the older cows go; they had had their day. I felt worse about the young cows and the heifers—young babes who never had a chance.

Beverly, his wife and farming partner, agreed: "It was the hardest thing we have ever done."

The two farmers described how they regarded individual animals:

> Sometimes cows went down after freshening. We'd figure out hoists to stand them up and hold them; we even made a "cow wheelchair." We've rolled them over on their side to milk, one and then the other. You'd maybe be further ahead culling them right there but you've got to try. It's heartbreaking how they'll do their best; drag themselves around, trying to get up. I've seen one be down twelve days and make it back to her feet.

These attitudes toward cows, based on deep knowledge and an affinity for the beasts, was typical of the craft farms and missing from the factory farms. As is the case in any process of production, with increased quantity there is less individual attention for any single unit in the production process. Since these "production units" are animals, this becomes a qualitatively different issue than the matter of making widgets.

Manure management highlights the nature of the craft farm. What to do with manure, given that it is an important resource when returned to the earth, is a perennial question on any dairy farm. It is natural, of course, to make manure handling as easy as possible. But the outcomes of decisions regarding manure have vast implications, and are among the issues that most fully distinguish the craft from the factory farms.

All but three of the thirty-four craft farms used solid manure systems (distributed with the "box spreader" pictured in figs. 122–27). The cows dropped their manure primarily in a trough behind their customary places in the barn; the farmer added soiled straw bedding and any spilled feed to the manure, making it solid enough to pretty much stay on a pitchfork. The straw, hay, and feed mixed into the manure soften its smell and make its consistency more like humus and less like animal waste. Most important, manure in this form returns nutrients more efficiently to the land because the more solid manure holds itself in place on the ground. The straw and other matter add to soil the substances that erosion takes away. What is called "solid manure," that is, manure mixed with roughage, is instant topsoil.

The solid manure system pictured in figures 122–27 has not changed a great deal since the SONJ era. Spreading the manure in the winter remained a cold but not terrible chore. In this system, however, there are benefits on several levels. The barn is cleaned daily, and the land is enhanced. For people living nearby, the rural idyll is preserved, because solid manure does not stink like liquid manure.

Structure and Culture

In general among decision makers in U.S. agriculture, the culture of the craft farmer has been seen primarily as an impediment to modernization and progress.[6] The establishment of cooperative extension and rural sociology departments has been predicated on the idea that farmers need to maximize their own human capital in the form of education, training, and the import of new ideas and techniques in order to overcome their limited and "premodern" cultures.

I observed two distinct cultures among the dairy farmers I surveyed in northern New York. The factory farmers epitomize the American ideal of rationality, efficiency, and profit maximization, embracing expansion and modernization as the twin pillars of their professional identity. Many became farmers as adults and from outside the region and farm to make a profit rather than participate in a lifestyle or extend a family tradition. This is, in general, the model of American agriculture promulgated in schools of agriculture and agricultural policy. The craft farmer, by contrast, extended traditions that may have been many generations old.

In spite of the reasonableness of the craft mode, several craft farmers, when asked in 1989 what farmer in the neighborhood they admired the most, cited the largest farmer in the neighborhood, then milking 268 cows. The sprawling cow factory seemed like an extraordinary new possibility to a farmer making do with craft ways and a herd of fifty to eighty milkers.

But other craft farmers were proud of being at the lowest end of the neighborhood social structure. One family, for example, identified themselves as distinctive precisely because of how much they did by hand. They grew most of the feed their cows ate and were content with a lower herd average. I was startled that they did not know about hormones to enhance milk production (then a topic of great debate in the neighborhood). They milked thirty-one cows with a bucket-and-transfer system, a system dating to the SONJ era. In general, their identity as farmers came from the simplicity of their systems and their relative independence from debt, fossil fuel, and machine lust.

The cultures of craft and factory farming also allocated work differently (see table 5). Women were more involved with production on the craft farms, assuming responsibility, on some farms, for all aspects of crop and cow work. On the factory farms, women's only contribution was bookkeeping. This is a natural conclusion, because the smaller dairy farms relied more heavily on family labor, and on the larger farms, most of the heavy tasks were done by hired men and the male

TABLE 5 Percentage of Various Jobs Performed by Women, 1989–90

	Craft Farms	Factory Farms
Paying bills	40 (14)	70 (9)
Keeping accounts	30 (10)	70 (9)
Monitoring births	18 (6)	8 (1)
Feeding calves	21 (7)	8 (1)
A.M. milking (primary)	6 (2)	0
A.M. milking (helping)	12 (4)	8 (1)
P.M. milking (primary)	15 (5)	0
P.M. milking (helping)	6 (2)	0
Cleaning milk machines	24 (8)	0
Feeding rations	9 (3)	0
Moving cattle	21 (7)	0
Field work	9 (3)	0

Source: 1989–90 St. Lawrence County survey.
Note: Numbers in parentheses are absolute numbers of farms.

principal operators. The implication of this insight is that on craft farms, women and men were more often full partners in the farm operation, and that each farmer, whether female or male, had the full range of knowledge of animals, crops, machines, and, of course, economics. The small size made specialization impossible, and it increased the challenge of farming, while certainly making it more interesting.

This pattern is reinforced by data reported in table 4, which indicates that craft farmers used less hired help per cow, often having no help at all. The hired men they did use were often members of their extended families, or they often became surrogate family after years on the farm. An older woman farmer described her knowledge of each cow and the pleasure she took monitoring them for health and breeding decisions. She asked rhetorically: "Is hired labor going to watch every cow to see if she is OK?" On the big farms, she said, "cows were treated like replaceable parts: they cull them for any little thing and replace them at the drop of hat." Some of her cows had been in the herd for a decade.

Thus the labor on the craft farms derived principally from the family itself, and that often included women as well as men. Whether it was for economic necessity or something else, the craft farms were more egalitarian on the matter of gender.

Labor Exchange in the Modern Era

While the formal labor exchange ended in the 1950s, some of the craft farmers still worked with their neighbors.

For example, a widow who had assumed responsibility for her farm after her husband died talked about how she had good animal intelligence but lacked mechanical aptitude and knowledge about cropping. Her neighbors had become her

Fig. 130 A farm with elements of both craft and factory farming, 1988. The farmer, a bachelor, had bought the farm from his parents in the 1950s. It was a neighborhood showpiece, admired for both the success of the Jersey herd (the smaller animals producing several thousand pounds per year less than Holsteins, but of of higher-butterfat and higher-protein milk) and the general sense of getting all the work done on time, neatly, and in an organized fashion. The farmer milked only fifty-five cows when this photo was made, and housed his cows in a stanchion barn, milking them through a pipeline system. However, he had built a small manure pit, visible behind the barn, which was well integrated into his overall plan.

The farm houses the cows in the original barn, to the upper right of the house. A second building next to the barn is a freestall barn, used as a holding area for cows. The farmer's equipment is neatly lined up, and much is stored in the Quonset hut, which even has skylights. Photograph by the author.

teachers and helpers. Without them, she said, she could not have made it. Charlie and the sons had showed her how to plow and how to plant. They were there when equipment broke. While I was visiting the farm for our interview, one of the neighbor's sons had stopped by to help unload bags of feed. He said he had seen the truck come up the road and figured he would lend a hand.

Only one farmer of those I interviewed wanted to recreate the cooperative arrangements of the past. He hoped to interest neighbors in shared ownership of machines such as round balers, which harvest hay much faster than old systems and thus could service several farms a season. His plans were built on inspiration from the changing works traditions he had participated in as a boy. In a sea of individualist farmers, he was distinctive. When I updated my data in 1999, I learned he had sold his herd. There had been no takers for his vision.

Indeed, I asked all farmers if they participated in informal exchange and only a few had a tale to tell. In one instance a farmer with large holdings lived next to a farmer with a much smaller farm. They became good friends and worked together for two seasons. The arrangement had ended when they could not agree on a definition of equivalence in contributions of labor and machinery. The bigger farmer supplied most of the equipment and felt the smaller farmer should work more hours on his larger fields than he would work on the craft farmer's landholdings. Though they approached the problem from several perspectives, they had not been able create an arrangement they saw as equitable. Their exchange of labor and equipment had ended, and, coincidentally, both of them had left farming by the end of the decade.

To draw these comments on labor exchange to a close, I recall the response of an elderly couple, from the only operating farm left in an old neighborhood along the St. Lawrence River. The farmer said,

> They are all gone-all those we changed works with. Out of business. Maybe a few are raising animals-we are the last dairy. You can't call on your neighbors to help you out-they don't even know how to drive a cow! It would be like if I was asked to assist a doctor in an operating room-I wouldn't know what to do. I might want to help, but I couldn't. That is the way it is now with the neighbors.

With qualifications as noted, the craft farms represented the logical extension of the SONJ era. The labor exchange of the earlier era had evolved out of existence, but the craft farmers had adapted their farms to several new technologies, which they had paid for with the gradually inflating price of milk. They grew most of their cows' food and used their manure to replenish their fields. They were farmers who were also neighbors, although they often lamented how "neighborliness" had declined. Above all, my visits to these farms confirmed their commitment to their animals as more symbiotic than exploitative. The cows needed the farmers, and the farmers were happy in their cow work. In the end, of course, both the craft and the factory farmers send them off to the slaughterhouse, but in the meantime there is a different quality in the relationships between animals and humans on the craft farms. Most of these farms were not in an economic crisis, and some were sufficiently profitable to support two generations of a family. I saw no need for this system to be dumped into the dustbin of history.[7]

Factory Farms and Megadairies

I classified twelve of the forty-six farms I studied in 1989–90 as factory farms. As can be seen in table 4, these farms were about twice as large as the craft farms. They had twice the land in production and three times the amount of corn. They had half the pasture per cow and more than three times the hired labor per cow. Herd averages are difficult measure, given the impact of mixed herds among both the craft and factory farmers. But when the herd averages are adjusted for these factors, the factory farmers averaged about a thousand pounds of milk per animal per year (about 8 percent) more than did the craft farmers. They kept their cows for about half the time of the craft farmer neighbors, and their calf mortality was nearly twice that of the craft farms. Nearly all the factory farms used free-stall barns, milking parlors, and liquid manure systems. But what is the flesh behind these numbers?

The system I have described as "craft" is a plateau against which habit and technology strain. This is especially the case in the United States, where tradition is generally of little use and capitalist mentalities saturate nearly all aspects of life. The bulk tank system, which came into existence in the 1950s, ended a system based on hauling milk in 50-gallon cans. If an employer of the milk company will drive to your farm to take your milk, a farmer might reasonably ask, why should one not want to increase one's production? Since the system is only indirectly affected by the market (since milk prices are governmentally determined), increased production theoretically leads to greater income, assuming costs can be controlled. It is this mentality which has destroyed the craft system.

Some farmers adopt the newest technology as a matter of course. They are more interested in machinery than animals or land. For these farmers, an operating rationality is rather simple. It means finding the best technology with which to maximize profit. Animals and land are inputs to the formula, no different than other inputs such as labor and machines.

My argument, as developed thus far, is that certain technologies and farming practices redefined farming, they did not just allow for farming to take place on a larger scale.[8]

The critical change is from a system using a stanchion barn and solid manure to a system using a free-stall barn and liquid manure. With this transition, the relationship between the animal and the farmer is transformed. I offer my fieldnote observation of a visit to a free-stall barn:

> After the stanchion barn, the free-stall is a polluted environment, owned by the cows. They wander about, expelling manure at random. Once a day a worker drives a small tractor with a blade through the barn pushing the manure into a trench which flows to a manure lagoon outside. The cows are al-

Fig. 131 A farm which fits the profile of the factory farm, 1988. The farmer milks more than two hundred cows, which are housed in the free-stall barn (the metal Quonset hut adjacent to the three large silos). The original barn, to the right of the free-stall barn, has been replaced with a shed that may, at this point, house dry cows or calves. The original silo, which held sufficient corn to feed a herd in the SONJ era, can be seen to the lower right of the shed. Its size compared to the new silos is evident. Across the street is the farmhouse and a trailer for the hired man to the right. The defining feature of the farm is the huge manure pit. Since the pits are only slightly more than head high, they are almost invisible from ground level and only show their presence through runoff pollution, smell, and spilled manure on roadways. Photograph by the author.

most never outside and don't seem too pleased to see me. If I were a farmer, I'd have to force myself to enter this inhospitable space.

At milking time, in the free-stall barn, the cows walk single file into an attached room called a milking parlor, where they stand in gated quarters waiting to be milked. Small parlors are "double-sixes," meaning two rows of six cows are milked at once. There are now parlors in which twice that number of cows can be milked. Milking is simple: a worker (note, I have not said *farmer*) attaches milking machines to an animal. When the milk is depleted from the udder, the machines detach on their own and swing out of the way. After the group of cows is milked, gates

Fig. 132 Inside a free-stall barn on a factory farm, 1990. Photograph by the author.

swing open and the group proceeds out of the parlor and back to the free-stall. Others jostle in to assume their places.

In the stanchion barn, the worker moves from animal to animal; in the milking parlor, the worker remains in one place while the work flows past. It is the introduction of the assembly-line process to what resembled a pre-assembly-line factory.

Milking in the parlor is efficient and unskilled. The worker's area is lower than the cows', so he or she is at eye level with the cows' udders. In fact, these workers describe cows as "teats and feet," which are the only parts of the animal they see or with which they have interaction. Figure 133 frames the animal as the worker sees it in the parlor.

The parlor de-skills the work of milking because it eliminates from the process of milking all actions but the single act of attaching the milking machine. The cows are fed automatically. There are, in fact, milking parlors just placed in service in which even this last human contact has been automated. These parlors are run by robots and monitored by computers.[9]

Cows feed themselves from bunkers in the free-stall barn. One farmer, who used a free-stall for his herd of nearly three hundred cows, volunteered that the free-stall barn was "harder on cows; timid cows are bullied by stronger cows and do not feed

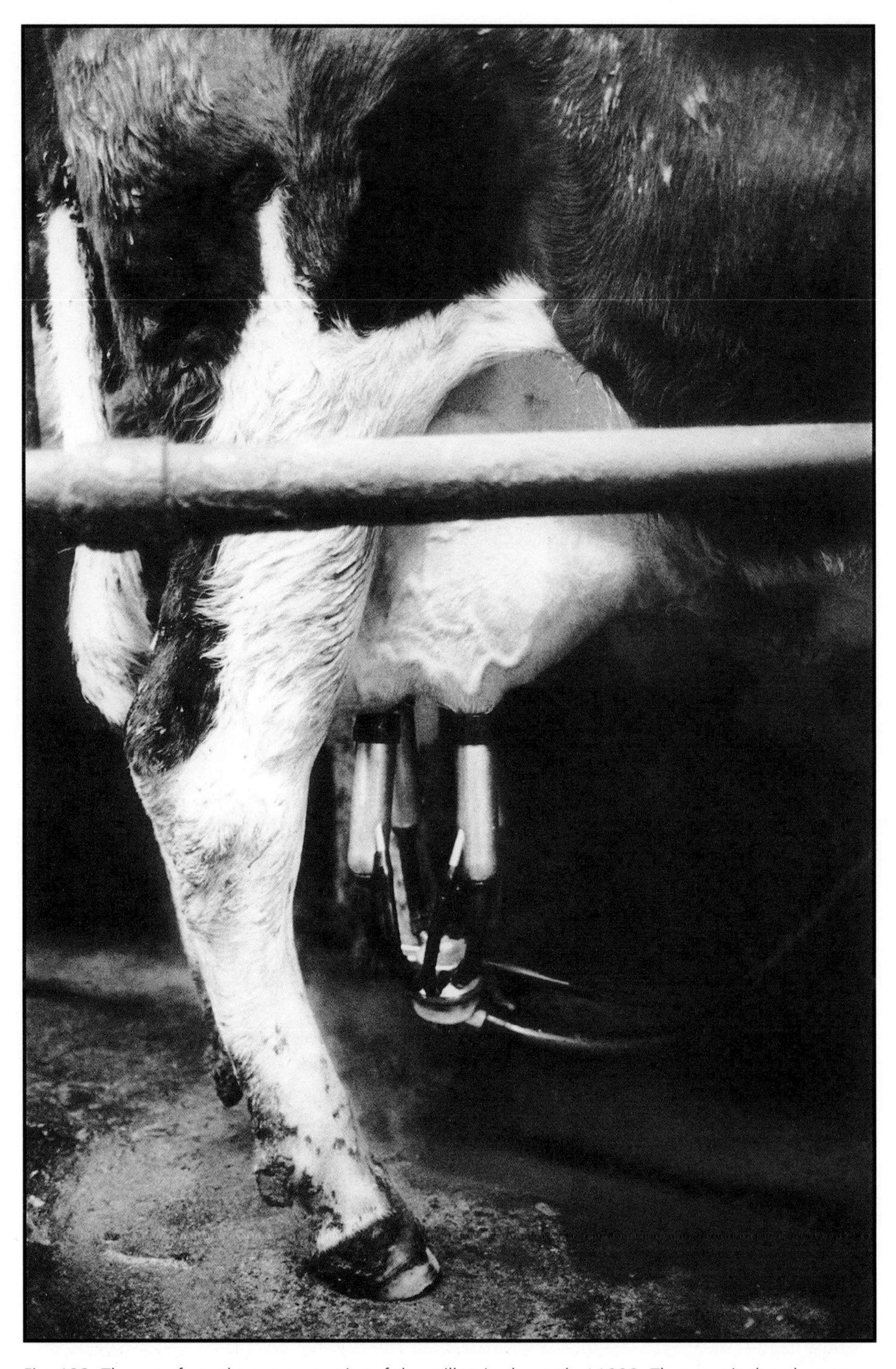

Fig. 133 The cow from the vantage point of the milker in the parlor, 1990. These agricultural assembly-line workers call cows "teats and feet." Photograph by the author.

well at the bunkers." Their feet are in worse shape because they continually stand in manure. On some farms, the area preserved for "dry cows" (those resting for a month or so after their annual milking sequence, before the birth of a calf) are not afforded the attention they receive on farms run more on the craft logic. We are reminded that calf mortality (see table 4) on factory farms is twice that experienced on calf farms.

Because cows are crowded together in an unsanitary setting, disease is easily introduced to the herd and spreads easily. Veterinary work is more difficult because individual cows are hard to locate, and because there is not usually an area separated from the barn itself in which to examine or treat animals. I've watched vets lance stomachs and do pregnancy tests in the middle of three hundred jostling beasts in the manury free-stall. Overall, while the stanchion barn puts the farmer in control of the situation, the free-stall belongs to the animals. Indeed, a person feels out of place in the space, even threatened. The cows become less used to humans, and the relationship between the two species loses empathetic character.

Manure is handled differently on factory farms. As herds increase, it becomes impractical to gather and spread the manure and bedding mixture. The alternative has been to scrape the soupy manure from the free-stall barn into a "lagoon" which stores the liquid until it is spread. On the surface, this seems like an entirely rational, though expensive, process. However, liquid manure introduces three problems: runoff into streams, groundwater pollution, and subjective environmental degradation, that is, smell and fouled roads.

I asked Jim Fisher, a farmer firmly entrenched in the craft mode, if the role of manure has changed as the free-stall system has become common. He answered:

> It surely has! Manure was always considered a good fertilizer and years ago it was always piled in the winter neatly. It was taken out with the horses in a wooden box. You drove the sleigh through the barn, and shoveled it in from each side. Two men went to the field and shoveled it into a nice, neat, square pile, and the efficient, careful farmers were the ones who had the nicest-looking manure piles. They were square and nice and didn't leach in all directions. Then, at the beginning of the season, in the early spring, the manure was spread quickly on the land when the most good of it would be gotten, before it was cultivated or planted . . . or in the fall before they plowed the corn ground. Then we went through a stage where I can remember in some of the farm magazines that manure was so much work and now that we had commercial fertilizer it would be better if manure was just put in a lagoon someplace and forgotten about, because the commercial fertilizer grew so much faster crops and it was so much more efficient. Farms were getting larger and they said that all this time it was taking to spread manure was really wasted

Figs. 134–35 The challenges of managing fields, 1989. Fig. 134: Flooded field. Fig. 135: Rutting land during a wet fall. Large farmers are less able to use land in a way that addresses the peculiarities of specific locations and conditions. Photographs by the author.

when you got right down to the dollars and cents of it. I read those things and I thought, "What are we coming to next?" But then, when we got into the energy crunch and the price of fertilizer went from a few dollars a ton to $150 a ton or more, manure took on a much pleasanter face again, and the agricultural economists began to think up ways to treat the manure as it should be. Now, of course, instead of piling it (which is impossible with the large herds) they are recommending the lagoon type of thing, or the harvester system, or some way of preserving the manure in pits or reasonably efficient places, and then once or twice a year getting it on the ground when it's most needed and is most efficient. It really cuts down on the price of fertilizer just tremendously now. Some folks say they hardly need to buy any fertilizer at all. So it has gone in quite a complete circle.

With the free-stall barns, where there is no bedding, it would be impossible to shovel it and handle it by hand. So it has been necessary to think up these other ways of taking care of it. These manure systems are also very expensive. But it's interesting to see how manure has finally come back into its own, more or less, as it should be. Apparently, it makes great crops but I certainly don't—

Emily: —Don't ask us!

Jim: I wake up at night sometimes and would like to write to Cornell and tell them what I think of that. I just can't imagine living close to such a stench! Some of our neighbors, when you go to visit them, if the winds are in the direction of the manure pit, or if they are spreading, which is probably thirty days out of the year-you just live in a horrendous stench. It's very maddening, especially to people who aren't in agriculture.

Doug: It is not a good smell—it's not like the manure from the barn.

Jim: Oh, no. It isn't a good smell at all.

Emily: We try to keep thinking-is it because we are not farming anymore that we mind the smell so much?

Jim: I don't think so. I like the smell of a lot of different manures-fresh manure-But when you go into some of these barns now—it is just sickening!

Farmers spread liquid manure with large tank spreaders, on a limited number of days of the year. The new huge herds of up to one thousand cows have created a surplus of manure, and the manure pit has become an environmental hazard. Recently a failed pump on a North Country farm caused a manure pit to empty back through a barn (the pit was above the barn) and to spill thousands of gallons of manure onto the adjoining land and road. A woodchuck dug a hole in the base of another farmer's pit, and the river of manure ran out of the lagoon and fouled nearby land and roads. The quantity of manure has led to field runoff and the pollution of streams, plentiful in the North Country. The manure has threatened ground water, particularly where soil is sandy. Finally, when farmers spread liquid manure they create a noxious odor which permeates the neighborhood. This stench is unprecedented in family dairy farming and ruins the pastoral idyll for farmers and others who share the country.

As I write in 1999, the state of New York has mandated that all farms using liquid manure systems have a plan in effect which will address the environmental and social problems of the liquid manure system within five years. While some problems can be solved with different technologies (for example, if the manure is injected into the ground, there will be less runoff) there are still no solutions for most of the crises caused by the several times greater production of manure on the largest farms. The simplest solution is to recognize that the megadairies create unsolvable problems and therefore to exercise public will to question their very existence.

The differences between craft and factory farms increase as the factory farms become larger. My argument is that the technology of the factory farm encourages the expansion of farms to several times what had been accepted as the natural upper limit. In the ten years since I completed my survey, this expansion has taken place within the state and the neighborhood, and as a result, there has been fundamental change in the neighborhood.

The Scene at the Turn of the Millennium

Table 6 indicates the fate, statewide, of farms of different sizes, between 1987 and 1997. All categories of farms up to two hundred cows have declined precipitously in ten years, and very large farms, with more than five hundred cows, have more than quadrupled in number. The total number of cows milked (not recorded in this table) has declined slightly. Thus, the cows have become aggregated in fewer and much larger farms.

These data must be considered in the context of the overall decline in agriculture (table 2): eight thousand farms at the turn of the twentieth century in St. Lawrence County, have become around six hundred at the present, farming about one-third of the land under cultivation at the turn of the century.

I have also determined how my sample of forty-six farms fared during the ten years since I completed my survey, which I present in table 7.

Of the thirty-four craft farms, nineteen (56 percent) remain in business in 1999

TABLE 6 Number of Dairy Farms in New York State, 1987–97, by Herd Size

Herd Size	1987	1992	1997	Change, 1987–97 (%)
1–9	1,623	1,068	777	–52
10–19	495	413	318	–36
20–49	4,728	3,340	2,351	–50
50–99	5,178	4,073	3,506	–32
100–199	1,492	1,389	1,210	–19
200–499	295	360	461	+56
500–999	29	47	88	+203
1,000+	NA	06	21	+250
Total	13,840	10,696	8,723	–37

Source: *New York Agricultural Statistics*, 1987–97.
Note: NA, not applicable.

and fifteen (44 percent) ceased their dairy operations during this ten-year period. This is a rather shocking place to begin, but the actions behind these patterns are not, in and of themselves, surprising.

Three of these fifteen farm families retired and did not pass their farm to the next generation. Four farmers sold their herds but continued to crop their land and sometimes to house other farmers' heifers or calves. They were farming land, but not milking cows.

Five of the fifteen craft farmers who left farming did so because of injury or death to one of the principle operators of the farm. One farmer was killed and another seriously injured in farm accidents; two died as elderly individuals, and two others were injured in nonagricultural accidents badly enough to force their retirement from farming.

Six other quit for various reasons: two farmers were newcomers who did not have the stamina or general acumen to remain in the business; four others left because of financial crises.

The farmers who left dairying ranged from the smallest to some in the middle range. At the time of the 1989 survey, those who were to go out of business averaged forty-one cows and 215 acres of tillable land.

The land on six of these fifteen farms passed out of agricultural production when the farmers sold their herds—in all, about 1,300 acres. Most of these farms were on dead-end roads or in areas of the county which were heavily wooded, with small fields tucked into woods. In an era of the expansion of factory farms, most land which passed out of production did so because the farms were isolated and hard for another farmer to utilize. At least one of the farmers who went out of business in a prime agricultural area continues to live in the house; his barn has fallen in and his fields are unused and returning to weeds. Several adjoining farmers covet his land and disdain his lack of interest in using it or having it farmed by others.

Nineteen of thirty-five farms (54 percent) which were classified as craft farms in

TABLE 7 Change on Forty-Six Dairy Farms, 1989-99

	Farms That Remained in Farming		Farms That Left Farming	
	Changes	No. of Farms	Changes	No. of Farms
Craft Farms (N=34)	Farms unchanged	11	Ceased farming due to death or injury of principal farmer	6
	Same farm methods; generational transfer (or in process)	4	Sold cows; continue auxillary, agricultural work	4
	Craft farm expanded to free-stall or milking parlor	4	Retired; land unused	3
			Quit farming; land unused	2
Factory farms (N=12)	Farm unchanged	6		
	Significant expansion	4	Quit farming; land absorbed by neighboring farms	2
	Total	29 (63%)		17 (37%)

Source: St. Lawrence County survey update.

Eleven of these operations are essentially as they were a decade ago, though two have modernized or built new stanchion barns. Four farms have not changed in size or organization, but they have transferred ownership to the next generation. Four previously craft farms have built free-stall barns, expanded their herds, and would, in 1999, be classified as factory farms.

What I called factory farms in 1989 (twelve of forty-six, or 26 percent) had become more factory-like by 1999. In the neighborhood I surveyed (including the immediately surrounding region), there are now three "megadairies," each milking between eight hundred and one thousand cows. One of these farms was, in 1989, the largest in my study, with 268 milking cows. This farm has tripled in size in ten years. The agribusiness company Agway has plans for a three hundred-acre facility for raising calves and heifers on the site of a previously successful craft farm. The development of that facility is temporarily on hold while the environmental impacts of the anticipated manure and other environmental effects are argued. If it is built, it will transform a beautiful corner, with one of the most well-run farms, into a rural factory.

Four of the twelve factory farms have experienced significant growth, as 250–300 cow herds become common in the North Country. This has changed the character of local farming, and it has affected the surrounding community.

The expansion of the largest farms has also added large truck and equipment traffic to the barely two-lane rural roads. Milk pickup for a thousand-cow herd is itself a significant addition to the traffic. As factory farmers buy the land previously belonging to their neighbors, their crop work becomes inordinately complicated.

Photograph by the author.

They must transport their tractors, plows, choppers, and wagons sometimes miles from one part of their farm to another. This adds to the caloric irrationality of the farms, as the fuel costs for heavy equipment are added to the caloric price of production. It also makes cropping inefficient and reduces the farmer's knowledge of his or her fields and lands. Finally, the increasing machinery on rural roads further alienates nonfarm residents.

The liquid manure of the megadairies has become an overriding problem, with symbolic as well as actual dimensions. What was a rural refuge has become, in the vicinity of the factory farms, a stinking and noisy industrial center.

An important way to seek the meaning of these changes is to place oneself into the experience or perspectives of the farmers, both those who left farming and those who remain.

Clearly, many of those who left farming did so for commonsense reasons. These include several who retired or were injured and, of course, those who died. In several cases, the farmers sold their herds but continue to grow crops and sometimes rent their empty barns to house their neighbor's heifers or calves. They have become, in fact, contract workers for the factory farms in the area.

Some craft and factory farmers were flushed out of the system because they were less competent or motivated, or less lucky than their neighbors. In some cases, their neighbors had been waiting for their seemingly inevitable demise for years. The percentage of failures of this kind is not surprising given the typical distribution of competence, motivation, and luck.

What is remarkable is that during these ten years, no small farms began in place of those who left. Most of the land on the small farms that ceased operation was purchased by larger farms. About a third of the land went out of agricultural production. These farms would be ripe for purchase by a new generation of small farmers. Interestingly enough, one of the only new small farms in the region (immediately adjacent to the area I studied) is run by a Vietnamese family that has gained the respect of neighbors unaccustomed to outsiders. With a great deal of family labor and attention to detail, this farm is becoming a neighborhood showcase, supporting a family with a small herd and limited land holdings. Just ten miles in different directions from the neighborhood I studied are two Amish neighborhoods, which continue to show that a nonmechanized farm, supported by informal collective work arrangements, can produce a viable agricultural system.

Yet these examples of viable small farms do not capture the imagination of most farmers. Many farmers see their fates, in one way or another, as tied to the factory farms and the new megadairies. There is spirited criticism of these trends by some farmers, resigned acceptance by others, and by some an enthusiastic endorsement that growth, even as represented by the megadairies, must be a good thing. At least some of the explanation for farmers' response has to do with the broader cultural embrace of entrepreneurial capitalism, the suspicion of government regulation, and our willingness, as a nation, to replace craft with industrial capitalism.

The largest farmers in my study did not report working any fewer hours than did the smaller farmers. Their debt loads were often more than a million dollars, and, while they had high cash flows, they were more vulnerable than smaller farmers to fluctuations in the price of milk. As they have come to depend on hired help, their problems of labor management have become one of their greatest concerns. Like all assembly-line work, the work in the milk parlors, placing mechanical pumps on an unending parade of udders, is not meaningful, interesting, or engaging. Not surprisingly, labor turnover on the megadairies has been a critical problem. And when workers do not appear, unmilked animals become very sick very soon. Given the scale of these operations, there is little room for the human or machine breakdowns which are themselves a function of the increase in the scale of production.

The biggest farms have not only absorbed the land of their neighboring farms, but also turned previously independent farmers into workers or subcontractors. As a result many rural families, previously independent, have come to depend on a single, huge farm which is financially and socially precarious. If the farm fails, it drags many others down with it.

The megadairies, and the generally increasing scale of dairy farming, threaten the very rural culture they are coming to dominate. Not all cultures, of course, are worth saving. I am sure many of us lament the end of the craft era just as we enjoy the benefits of a mass-production world. But in farming, it is different. Milk does not have to be made in a factory. In fact, there are no benefits in doing it this way, and there are enormous costs. The costs are experienced by the animals, the land, and nonfarm neighbors, but they are also shared by the farmers I met who had become harried and driven by the demands of an incredibly complicated system. The pleasure had gone out of farming. Their cash flow is extraordinary, and perhaps they feel rich. But with milk fluctuating between twelve and seventeen dollars a hundredweight in the six-month period preceding the completion of this book, the fearfulness of their own vulnerability must become nightmarish. The payments on their debt did not fluctuate with the price of milk.

To end this consideration of the "changing works" of American dairy farming, I look briefly at dairy farms in two other modern, industrial nations—Switzerland and France.[10] It is, of course difficult to compare the farm systems in these countries, given the different national cultures and practices.[11] But much of American agriculture is derived from European immigrant practices and has evolved in the context of American capitalism, a seemingly uncompromising growth ethic, and abundant land and low population density. Yet we are basically similar to farmers in Europe: the cows are the same, the machines similar, and most of the crops are cousins of our own. Thus, perhaps a glance at some of their alternatives is worthwhile.

For several reasons these national cultures are more sympathetic to the collective orientation generally, and social practices similar to what we have called "changing works" have not been lost in these two systems. In Switzerland, the farmers I interviewed purchase equipment together, such as tractors, and use it as co-owners. Most of the more expensive machinery is owned by a co-op and rented or used by co-op members during the brief periods of harvest. It is the same in France, where the co-op organization is extensive and many layered. Co-ops with several thousand members have a primarily political mandate. Other smaller co-ops own equipment, which individual farmers rent. Much smaller co-ops are working groups among contiguous farms. A farmer might, for example, rent equipment such as a round baler or a thresher from a co-op, and several neighbors would gather to do the harvest on one farm after another.

What caught my ear, however, was statements made by a French farmer, an agricultural official, and an agricultural professional in a lengthy interview regarding farm culture itself. Each of them spoke of the co-op system as a necessary antidote to individualism. Farming, they asserted, is by its nature collective and community based. Policies must be formed which work against what they saw as pernicious individualism. Thus, the farmers understood the allotment system and government pricing as necessary given the vulnerabilities farmers faced from within and without. They could make their farms more profitable by making their milk more efficiently and with higher quality (in France measured primarily by a sanitary standard and protein content), but they would not be allowed, as in the United States, to increase from a hundred milkers to a thousand. The impact on the region and on the culture of farmers would be considered, by the body politic, more important than the ability of single farmers to grow as large as their vision would inspire.

The dairy systems in these and other countries, such as Canada, offer reasonable models. Only now, as rural people in the United States begin to organize against the pollution caused by large hog and chicken operations, have nonfarmers begun to enter the dialogue about the design of an agriculture which will serve the needs of farmers, the environment, rural communities, and nonfarmers. We chafe at the idea of government regulation and live with the rough world that relative lack of regulation creates. It is startling to look at another system in France, nominally like ours, and see prosperous farms with herds of forty milkers, a gorgeous landscape, and the most exquisite dairy food in the world. Perhaps we should look a bit harder.

So I end with this question. What is it to farm? Aldo Leopold wrote in the 1940s:

> There is as yet no ethic dealing with man's relation to land and to the animals and plants which grow upon it. . . . The land-relation is still strictly economic, entailing privileges, but not obligations.
>
> The extension of ethics to this third element . . . is, if I read the evidence correctly, an evolutionary possibility and an ecological necessity.[12]

A land ethic, indeed. But Jan Wojcik put it even more simply: "how should farmers farm? From time immemorial there has been really one answer. Farmers should farm so that they can farm again."[13]

On the surface, this seems an easy mandate; in the context of history, and considering the ethical considerations due animals and land, the needs of farmers, and the rights of citizens, the question becomes much more complex.

Notes

INTRODUCTION

1. This region, between the St. Lawrence River to the north and the Adirondack Mountains to the south, is referred to as the "North Country" so commonly that it is accepted as an unofficial name of the region. I realize there are several "north countries" in the United States, but no others, that I know of, which call themselves that.

2. I was ably assisted during this stage of my research by Karen Kline, who did several interviews and helped me with data analysis.

3. Plattner's research is published in his 1983 book *Roy Stryker: U.S.A., 1943–1950: The Standard Oil (New Jersey) Photography Project.* In the late 1970s Plattner interviewed most of the SONJ photographers and several Standard Oil (now EXXON) employees involved with the project. Plattner's research is particularly useful because many of the individuals involved are now elderly or deceased.

4. Plattner 1983, 18

5. Plattner 1983, 21.

6. Three influential books were produced during the era using FSA photographs. Essayist Erskine Caldwell and photographer Margaret Bourke-White published stories and photographs gathered during the summer of 1936 from nine southern states in *You Have Seen Their Faces* (1937). *Let Us Now Praise Famous Men,* by the documentary essayist James Agee and the photographer Walker Evans (1939), focused on the experiences of three white sharecropper families in Alabama. Photographer Dorothea Lange and economist Paul S. Taylor described the interlocking economic, social, and political forces causing agricultural migrancy during the depression. Lange felt that their book, *An American Exodus: A Record of Human Erosion* (1939), resembled, in intent and form, the photography and anthropology of Bateson and Mead (Ohrn 1980, 112), but Lange and Taylor's book to be yet another unrealized opportunity for the emergence of a visual sociology based on the documentary tradition.

The FSA photographs have been reprinted in many contemporary collections. Several studies stand out: Fleischhauer and Brannan's collection (1980) contains photographic projects from fifteen FSA photographers, including contextual information about the images. The book includes an essay by Alan Trachtenberg which evaluates and balances the questions of the ideological character of the FSA photography. Curtis (1989) also analyzed the ideological contexts and cultural backgrounds of several FSA photos and essays. Finally, Bill Ganzel (1984) rephotographed many of the FSA sites and people in startling fifty-year comparisons. This list only scratches the surface of the scholarship devoted to the FSA project. Individual photographs, such as Dorothea Lange's *Migrant Mother*, have become icons in the documentary movement. In many people's eye, the FSA project represents documentary history itself, though its hallowed place in the history of photography has been, as noted above, placed by several theorists in the context of its social construction.

7. Nearly thirty photographers worked on the SONJ project at one time or another. The most well known include Gordon Parks, Berenice Abbott, Charlotte Brooks, Esther Bubley, Harold Corsini, Elliott Erwitt, Sol Libsohn, Lisette Model, Martha McMillan, Louise Rosskam, Charles Rotkin, and Todd Webb. Several of the photographers had worked for Stryker in the FSA, including John Collier, who later published the first book on visual anthropology (Collier 1967; revised and expanded with his son Malcolm, 1986), Russell Lee, Gordon Parks, Edwin Rosskam, and John Vachon.

Among the future careers of importance which were born during the SONJ project was that of Gordon Parks, who later became a *Life* magazine photographer, filmmaker, poet, and essayist—a leading African American cultural figure of the late decades of the twentieth century.

8. Stryker's idea of "shooting scripts" developed through his discussions with Robert and Helen Lynd, sociologists who, during the 1930s, wrote two influential studies of American communities. It was during these discussions that the notion of "visual ethnography" evolved, at least for these thinkers.

Stryker encouraged a documentary photography that differed from the common fine arts treatments of industrial society. Fine arts photographers, having little knowledge or interest in human work and industrial processes, used the industrial landscape as an artistic subject. Plattner refers to photographs such as Stieglitz's " Hand of Man," a murky, beautiful rendering of a train in a trainyard, made in 1902. More contemporary to Stryker, Minor White (about to become the first editor of the most important journal of fine arts photography, *Aperture*) made many industrial landscapes.

9. Plattner 1983, 16–17.

10. Plattner 1983, 11.

11. Plattner 1983, 12.

12. Plattner 1983, 21.

13. The SONJ northeast dairy farm photographs are remarkably consistent in style and content. The style is partly due to cameras used by the photographers. These were tripod-mounted view cameras, hand-held 4 x 5 inch negative press cameras, and 2 1/4 x 2 1/4 negative double-lens roll-film cameras. The view cameras were hard to move and thus used for landscapes and portraits; the press cameras, used by photojournalists during the 1930s and 40s, had to be loaded with a sheet of film each time an exposure was made, and thus they imposed a severe discipline on the photographers; and the roll-film cameras, awkward by today's standards, were the only cameras the photographers could carry comfortably. The roll-film cameras allowed photographers to photograph social life as it carried on around them, at least for twelve frames. None of these cameras used a wide-angle lens. The roll-film camera's lens was 80 mm, which is about what one eye sees. Thus, the photographs focus on a narrowly framed subject, which is often in the center of the frame. The result is that the photographs communicate an unambiguous theme. These characteristics of what has been called "modernist documentary" have been the frequent target of postmodernists, who identify the style with a sociological naïveté. See, for example, the essays collected by Richard Bolton (1989).

The endorsement of technologically advanced farming is implied in most of the images used in my study. We see hay harvesting with a crew loading loose hay, and with a baling system that was soon to make the loose hay obsolete. We see corn harvested with a changing works crew, and also with the first generation of farm choppers. Milk is still delivered to processing plants in 10-gallon cans, but we see hints of the future in the bulk trucks which would soon redefine the milk distribution system and the farm itself. Farmers still cut wood together, but soon they will heat their homes with oil. A few farmers still use horses, but we see their replacements in tractors and other machines. Harness makers and blacksmiths seem plucked from the nineteenth century. At the state fair farmers confront the new machinery that represents the future of mechanized farming. Sometimes they look perplexed, sometimes deeply interested. They certainly do not look bored (though their wives occasionally do!). Finally, the new technology is mated to science.

The SONJ photos do not show problematic aspects of rural life. The people represented in these images seem well integrated into their social roles. The images communicate the "virtues of farm life": hard work, inventiveness, a healthy balance of individuality and collectivity. The facial expressions recorded by the SONJ photographs suggest people comfortable in their lives—decent people working hard. This is a rural society in which parts fit together well, a society portrayed as a big team of well-adjusted players.

Social institutions and organizational arrangements contribute positively to the social lives of the farmers. Brooks shows us a harvest festival where, in 1945, the church distributes food to the less well-off. The state fair is a cornucopia of technological wonders; there are no photos of the banks or government loan offices which will organize the process that will put most of the farmers out of business and the rest into debt to purchase this equipment. The milk distribution is organized by the invisible hand of the market; farmers squirt it from their cows and deliver it in 10-gallon cans to a distribution center; others transport the processed milk to the city, meeting the needs of an industrializing nation. The only images of capitalist exchange have buyers and sellers of cattle meeting in public to exchange their animals to each other's advantage. In fact, it is such an important event that many farmers have come as spectators.

Most important, the photographs do not (nor could they) communicate the structural transformation that underscored the technological change occurring at that time. That a majority of the farmers pictured would be out of business as the technology was adopted is beyond the portrait of seemingly resilient and hard-working people.

14. Lemann's *Out of the Forties* (1983) is a collection of SONJ photographs and interviews with subjects of the photographs he located in the early 1980s. Okrent's collection (1989) follows a similar format. The author located and interviewed New Englanders photographed by the SONJ photographers. His book reflects upon the changes in New England as interpreted by the SONJ subjects. He writes: "It's no surprise that the past looks prettiest to those who haven't lived through it . . . our imaginings are far prettier than the more accurate memories of people a generation older" (19). The images chosen for the covers of these books, however, are clichés of rural bliss. Okrent uses Charlotte Brooks's image of four children running joyfully from the door of their one-room schoolhouse; Lemann uses Rosskamm's image of a nine- or ten-year-old boy in a sailor hat and belt spitting out a watermelon seed after tackling the largest piece of watermelon imaginable. Both of these projects provide a "generalist's cultural history," though Okrent's book has more depth and detail. Okrent has also written captions which update the circumstances portrayed in the images.

Ulrich Keller's 1985 collection focuses on the "highway as habitat," examining the rapidly changing road and highway system on an America moving quickly from rural to suburban.

My project does not overlap these books, but there is information in Okrent's collection that fills out an important part of my story. The photograph I use of Sylvester Moore leading his bull (fig. 104) presents him in a way that leads a viewer to assume rural solidness and respectability. Okrent tells us that Sylvester was an alcoholic who subsequently passed the farm to his son Turney Moore, who did not make a successful transition to the new era; he stopped farming in the 1950s and sold house lots from his road frontage. Moore eventually sold his house and moved into a house trailer on a small corner of the original farm. His son became a professor and left the region. The fields that two generations of Moores farmed are now brush.

15. DeVeaux 1997, 2.

16. Collier 1967. Some contemporary examples are van der Does et al. 1992; Gold 1991; Suchar 1988; and Harper 1987.

17. Liebow's 1993 ethnography, more than any other of which I am aware, presents the people he wrote about with their humanness showing.

18. Fisher's 1985 memoir was written to preserve the family history. It is, as well, an excellent repository of historical details and insights.

19. Lee n.d.

20. Harper 1987. There are parts of this book which overlap the earlier one, *Working Knowledge.* Willie's discussion of the methods of the blacksmith in this volume, for example, will be familiar to readers of the earlier book. I want to note, moreover, the extraordinary pleasure of working with Willie on the second book, which continued until nearly the moment of publication. Our last interviews and discussions about this project took place in October 1999.

21. Bateson and Mead 1942.

22. See Blumer's essay "The Methodological Position of Symbolic Interactionism" (1969).

23. Howard S. Becker's definition of culture (1986) remains both inclusive and helpful to a researcher making sense of a setting.

24. Sloan 1965.

25. This history draws from McMurry's superb 1995 study of early-nineteenth-century dairy farming in Central New York.

26. The nineteenth-century agricultural census listed production on each farm. For example, a typical farm of 170 acres in 1865–66 is recorded as producing 25 tons of hay and 2 bushels of grass seed, 700 bushels of oats, 100 bushels of corn, 50 bushels of apples, 1,100 pounds of butter (with thirteen cows over one year old), and 125 pounds of maple molasses (from which vinegar was made). The farm census also reports poultry, sheep, swine, and other cash crops. Not recorded is production purely for home consumption (St. Lawrence Historical Society 1985).

27. McMurry 1995, 40.

28. The statistics come from "The Case for Vegetarianism," lecture by Howard Lyman, Boulder, Colorado, 22 January 1997; Perelman 1976. The theme of energy use and industrialization of agriculture are discussed in several of Wendell Berry's essays in *The Gift of Good Land* (1981).

29. Weitzman 1991, 60.

30. Canine 1995, 47.

31. Canine 1995, 21.

32. The best photographs I have seen of this process are by Ernst Brunner, a Swiss photographer, who photographed hand winnowing in Switzerland during the 1940s. See Brunner 1996, 152–157.

33. Laxness's 1930s novel, set in his native Iceland, was translated and recently republished in English. Beat Sterchi's novel *Cow* (1988) offers the perspective of a Spanish im-

migrant who manages the animals on a modern small dairy in Switzerland—a portrait from the vantage point of the animal as well as the man.

34. McMurry 1995, 12.

THE MACHINE IN THE GARDEN

1. The European context for this issue was the transition of feudalism to industrial capitalism. The loss of the old world meant also the loss of rigid stratification systems based on ascribed status. In the United States the lack of a feudal past led to the idealization of the "yeoman farmer," who seemingly would combine independence of thought and action, maximizing both production and human potential.

2. Harper 1991.

3. There is, however, growing interest in human-animal interaction, as indicated, for example, in the recent special issue of *Qualitative Sociology* (vol. 30, 1994) dedicated to that topic. In this issue, however, most emphasis is on the treatment of animals by people: the alienated environment in which most people experience animals, social movements which mobilize humans to treat animals more "humanely," and the like. The sense of the animal as an actor in a symbolic interactionist environment does not enter the picture.

4. Blythe 1969, 54.

5. The following discussion relies heavily on Canine's 1995 study of how a small team of inventors and entrepreneurs redesigned the field combine. Canine's study includes histories of several agricultural machines.

6. Canine 1995, 158.

7. Morrissey's (n.d.) memoir is a delightful read, well illustrated with photos from his several decades as a tractor salesman. For Morrissey, the reluctance of farmers to mechanize was his principal frustration. His job was to overcome it.

8. Hostetler 1980, 126.

9. Hostetler 1980, 127.

10. "Nitrogen fixing" is well described by James Trefil (1997) in *Smithsonian* magazine (one might also consult any number of biology textbooks). Nitrogen, essential to life, composes 80% of the air, yet it is unusable in its naturally occurring form as nitrogen molecules, which each consist of two nitrogen atoms. To be used by plants, the nitrogen molecule must be separated and the atoms attached to other atoms, such as oxygen. In nature this process, called "fixing," is done primarily by bacteria, many of which live in nodules on the roots of plants called legumes. For centuries farmers have known that planting legumes, such as soybeans, would rejuvenate worn-out fields by returning nitrogen to the soil. Trefil describes the environmental dangers of overuse of manufactured nitrogen fertilizers, now at about eighty million tons a year worldwide. This nitrogen increases the global supply of fixed nitrogen. In the form of runoff from fields or lawns, or carried by the air, it overfertilizes streams, lakes, or bays, changing ecosystems with implications for all aspects of the food chain. Trefil writes that "in only fifty years we have come to dominate the nitrogen cycle. In some parts of the world, notably Western Europe, there is the potential for 'nitrogen saturation,' in which the ecosystem is storing all the nitrogen it can, and any more . . . may be transported into the air or water" (76). Overnitrogenated land produces more biomass but less diversity in plants. This might be scum in lakes or airborne nitrogen, which produces haze. The implications of oversaturation of nitrogen are profound but only now moving to the forefront of environmental consciousness.

11. Sloan 1965, 94.

12. Several scholars have studied how agricultural labor exchanges are affected by forces and manifestations of the modern era. For example, Spencer (1979) studied how minimum wage legislation affected traditional patterns of labor reciprocity in Sierra Leone; Guillet (1980) showed how capitalist influence in Andean peasant farming did not eliminate the im-

portant role of reciprocal labor exchange in agricultural work. Other scholars have noted the important role labor exchange continues to play in diverse subsistence activities. Ingold (1983) studied the labor cooperation that made reindeer management in a northern Finnish rural community possible; O'Neil (1982) detailed the egalitarian character of cooperative work practices during corn harvesting and threshing in rural Portugal, and Gunasinghe (1976) studied the social organization of reciprocal labor exchange in Sri Lankan rice harvesting.

13. Durkheim's analysis of what he called the "common conscience" is described in *The Division of Labor in Society* (1933), 287.

14. Bush, 1987, 221.

15. Anderson 1940. Melvyn Dubofsky's history of the I.W.W. (1969) describes the politicization of agricultural migrants.

16. Jellison 1993, 109–110.

17. Rendleman, 1996, 82.

18. Canine (1995, 159–163) tells the story of the impact of the war on agriculture experienced by his family.

19. Klinkenborg 1986, 144–146.

20. Canine 1995, 125–145.

21. Stadtfeld 1972, 175–194.

22. Klinkenborg 1986, 141.

23. Canine 1995, 88–92.

24. Canine 1995, 128–145.

25. Adams 1994, 66.

26. Carolyn Sachs (1983) applied the concept of patriarchy to women and agriculture: "Understanding the position of women on farms requires an understanding both of the economic forces operating upon the structure of agriculture in particular and of the position of women in society at large. . . . A patriarchal division of labor operates in several ways. First, men rarely perform women's work. Since men's work is more important, they do not have to be involved in domestic work. Second, men attempt to control their own realm through the exclusion of women. . . . many farmers are caught in a cost-price squeeze and are unable to keep women removed from agricultural production. . . . In many instances, women do work in the male realm of agricultural production. In these instances, men decide which work women perform (xii).

Many feminist scholars have developed research from this lead, especially Fink (1986, 202–212), in relationship to farming in northeastern Iowa, and Meyer and Lobao (1997, 207) described how the recent agricultural crisis affected families: "most women experience farming as 'farm wives,' in structurally subordinate positions in the family enterprise. This position confers specific roles and responsibilities, subject to human agency, in farm production and household reproduction.

"Women are situated in a gender-delineated labor process in which the timing, nature, and outcomes of their farm work tend to be directed by men and coordinated around household exigencies. Although the gender division of farm labor varies by the characteristics of the enterprise and the commodity, family norms and ethnicity and other factors, the general trend is that men retain greater control over management decisions, such as marketing and input purchases, and major production tasks, such as field work."

27. For historical antecedents, see especially Nancy Grey Osterud's chapter on "work relationships among women" (1991, 187–201). Deborah Fink, describing gender roles in central Nebraska, made the observation "When, for example a farm woman went to her neighbor's home to help her cook food for a threshing crew that included both of their husbands, they were forming a wider rural social network on the strength of their roles as farm wives" (1992, 6). Bush described the extensive exchange of women's labor to prepare harvest meals for the crews of twenty to thirty men harvesting wheat in the American West during the

1930s: "Just as neighboring farmers combined and pooled horses and equipment, farm women often pooled their pots, pans and labor. A woman would be in charge of the cooking operation while the workers were harvesting on her farm. When they left, she would often go to a neighbor's home to help cook the meals for the harvest crew there" (1987, 218).

28. Jellison 1993, 29.

29. Jellison (1993, 176) describes the travails of a woman from Denver who married a Wisconsin farmer in the war years: "During the first year of her marriage, when wartime shortages prevented her husband from purchasing a modern combine or mechanical corn picker, McKnight had to cook for large numbers of threshers and corn shredders. She realized her cooking talents did not live up to those of her neighbors, and she resented the hard work. And because she was a newcomer, none of the neighborhood women offered to help her feed the hungry field workers." These events encouraged her husband to modernize his field operations, removing him from the labor exchange.

30. The idea of a rational basis for social exchange is as old as classical economists, notably Adam Smith, who applied the metaphor of the market to social exchange. This perspective is based on the assumption that humans act rationally to advance their own self interest. Modern sociologists, such as Homans (1958) and Blau (1964), can be considered proponents of the rationalist perspective, though their accounts of interaction in small groups stressed how differing power influenced exchange. Blau suggested that these differences in power became institutionalized into social structures, thus explaining the emergence of the social order. Blau used game theory to analyze balance sheets, which must underlie the profit and loss accounts of rational social exchange.

31. The cultural explanation identifies social exchange as embedded in rituals indicating identity, social role, status, and community membership. Malinowski (1922) asserted that people enter into exchange out of the mutual obligations that shape their interlocking status duties. Anthropologists such as Mauss (19670, Sahlins (1972), and Firth (1967) have argued that the character of non-Western society—interactively rich—leads to patterns of exchange that are categorically different from those experienced by the rational, self-interested modern individual. Thus, anthropologists do not so much deny the economic model of human motivation as add their own model of non-Western societies.

32. Reacting to Spencer's utilitarian assumption that human cooperation consisted of rationally motivated exchange, Durkheim proposed that decreasing homogeneity and increasing division of labor (modern, or "organic," society, changed the basis of social solidarity. In Durkheim's view, "if the division of labor produces solidarity, it is not only because it makes each individual an exchangist, as the economists say; it is because it creates among men an entire system of rights and duties which link them together in a durable way" (Durkheim 1933, 406).

33. A recent essay by Davis (1992) offers an alternative to economic or anthropological approaches grounded on rational choice and self interest or entirely ritual explanations of behavior. Davis emphasizes the moral and symbolic order and concludes that the consequences of exchange are not in the working of economic laws, "but in the social rules of power, symbol, convention, etiquette, ritual, role and status" (1992:8).

34. Gouldner's (1960) argument is that the norm of reciprocity is universal in human life. Gouldner points to the important contribution of reciprocal arrangements for creating social stability, while noting that many exchanges are not reciprocal, especially when large power differentials exist between the parties, or when custom or culture sanctions unequal exchanges.

35. The works I have relied on include Deborah Fink's 1986 study of contemporary farming in northwestern Iowa (*Open Country, Iowa: Rural Women, Tradition and Change*) and her 1992 study of gender roles in the Great Plains (*Agrarian Women: Wives and Mothers in Rural Nebraska, 1880–1940*); Jane Adams's 1994 study of southern Illinois (*The Transformation of*

Rural Life) includes the settlement to the present. Adams's scholarship includes several papers that develop arguments in her book (1988, 1993), as well as her edited memoir of Edith Bradley Rendleman (1996), a farm woman from the region she studied who was born at the beginning of the century. Nancy Grey Osterud's 1991 study of dairy farmers in the Northeast (*Bonds of Community: The lives of Farm Women in Nineteenth-Century New York*) constructs history through personal accounts as well as quantitative data. Two shorter case studies, found in the 1987 collection edited by Carol Groneman and Mary Beth Norton, are Corlann Gee Bush's analysis of gender relations during the mechanization of wheat farming in Idaho and Sara Elbert's study of dairy farmers just south of those I studied in New York.

36. Sachs 1983 is an often cited introduction to patriarchy in agriculture.

37. Fink 1992, 33.

38. This life (actually a bit to the east of the specific region Fink describes) was portrayed in Ole E. Rölvaag's *Giants in the Earth* (1927) written early in the era Fink describes. In Rölvaag's view, the severity of the existence led to madness, suicide, and social breakdown.

39. Fink 1992, 43.

40. Fink 1986.

41. See, for example, Fink 1986, 62 and 149; and Adams 1994, 88.

42. Jellison 1993, 157; Fink 1986, 146.

43. Adams 1994, 3.

44. Adams 1994, 89.

45. Rendleman 1996.

46. Adams 1994, 103.

47. Osterud 1991, 150.

48. Osterud 1991, 150.

49. Osterud 1991, 156.

50. Osterud 1991, 155.

51. Salamon 1992, 119.

52. Salamon 1992, chapter 5, especially table 5.1, 124.

53. Osterud 1991, 283–4.

54. Osterud 1991, 284.

55. Osterud 1991, 285.

56. Ulrich 1987, 37.

57. In but one example, Jellison cites a 1915 survey: "Midwestern farm women seemed to complain chiefly that their work was under mechanized. At a time when men increasingly employed steam- or gasoline-powered equipment, including the newly available gasoline tractor, most Midwestern farm women relied on their own physical labor to perform their daily tasks" (1993, 11).

58. Jellison's report on how women were encouraged to assume men's mechanized tasks—and how they did it—includes a 1942 advertising poster featuring a "tractorette": a beaming woman cultivating a field from the seat of her tractor.

59. These messages were promulgated in advertisements in farm magazines, and in booklets and pamphlets published by agricultural machinery companies, and by the extension service itself.

60. Jellison (1993, 23) tells us that during the 1920s, when most farms reported a profit of around $2,000, self-installed plumbing cost in the range of $300, a home heating system $200, a 10-horsepower gasoline engine (for running an electrical generator and other uses) cost $230, while a radio set a family back $175, nearly 9 percent of their yearly disposable income!

61. Stadtfeld 1972, 50.

62. There has been a great deal of discussion of how women and men see, and thus photograph, differently, but nearly all scholarship has dealt with the photographer as artist rather than as documentarian. Two early studies of female photographic art are the 1975

collection *The Woman's Eye,* edited by Anne Tucker, and Joyce Cohen's 1978 collection of self-portraits by women, *In/Sights.* The themes of self-reflection, self-portraiture, personal exploration, and development have been the focus of hundreds of essays, exhibitions and books. Recent examples are the 1995 collection edited by Neumaier, the 1991 collection on lesbian photography edited by Boffin and Fraser, and Abigail Solomon-Godeau's 1991 discussion of Francesca Woodman's photographs.

Few scholars, feminist or otherwise, have studied how women and men see the documentable world differently. Deborah Bright's 1989 study of women landscape photographers is a notable exception. Bright questions the assumptions behind photographers such as Linda Connor, who suggest that women photograph landscape differently than men because women are by nature less acquisitive, manipulative, and territorial than men, and she shows how such assumptions are ironically used to "devalue women and their cultural production" (142).

Several historians contributed to a project eventually authored by Diane Galusha (1994), which presents the work of three female photographers (Anna Carrol, Edna Benedict, and Lena Underwood) who lived in south central New York in the late nineteenth and through much of the twentieth century. The photographs collected in the volume appear to offer a female "sense." Many are family portraits that document growing children. Many of these photographs are taken at the eye level (or below) of the children, which we may interpret as indicating a higher level of empathy than in adults who look at and photograph children from their elevated level. Several images detail farmhouse interiors. The few photographs featuring dairy farming suggest a woman's view: the plate reproduced on the cover shows a child milking a cow, studious and competent. Anna Carrol's image reproduced on page 141 shows two young women feeding the family flock of chickens. It is a remarkable photograph that shows women in fashionable and long dresses in a typically dirty chicken pen; in the long exposure most chickens are blurred but the women are still and composed, one kneeling and the other standing behind holding an uncooperative hen. While the collection, published by a local historical society, is limited in its scope, it offers a view of dairy farm life that differs substantially from the SONJ images, done mostly by men.

An interesting project for the future would be to compare men's and women's contribution to an entire documentary project. The SONJ archive in its entirety would offer fodder for such a project.

63. Canine 1995, 96.

64. Canine 1995, 102.

65. Cohen 1992, 82.

66. Cohen 1992, 122.

67. Cohen 1992, 124.

68. McMurry 1995, 24.

69. McMurry 1995, 58.

70. Other methods of collecting semen are described in Stish Chandra Dasgupta's *Cow in India* (1945, 953). This comprehensive and poetic book describes the state of dairy farming on the other side of the globe, in a society about as different as one could imagine, during the same era. Amazingly, however, the similarities in the two dairy farm systems are more striking than are the differences, once one gets used to the idea that the Indians did not eat their culls.

71. Dasgupta 1945, 890–892.

THE HISTORY SINCE THEN

1. Karen Kline, who worked with me on the project during this era, and I interviewed forty-eight farmers, but used data from forty-six farms, two having been "farms" only in a nominal sense.

2. James Sterngold, writing in the *New York Times* (1999) reports that dairy farming in the Chino Valley in San Bernardino County, California (forty miles from downtown Los Angeles), concentrates 25 cows per acre on farms that average 880 animals. The farm described by the author milks more than 2,000 cows, which are housed on two hundred acres. These farms are, in the words of the author, "operated with machine-like precision and a pure business philosophy."

3. Gerth and Mills note that Weber "felt that social scientists had the choice of using logically controlled and unambiguous conceptions, which are thus more removed from historical reality, or of using less precise concepts, which are more closely geared to the empirical world. Weber's interest in world-wide comparisons led him to consider extreme and 'pure cases.' These cases became 'crucial instances' and controlled the level of abstraction that he used in connection with any particular problem. The real meat of history would usually fall in between such extreme types" (1946, 59–60).

4. Glaser and Strauss (1967) called this process the development of "grounded theory." Howard Becker's recent *Tricks of the Trade* (1998) describes the process of how a researcher defines concepts from fieldwork observation. In my case, at first all dairy farms seemed much more similar than dissimilar but the accumulating knowledge during fieldwork led to concepts that were subsequently reinforced. Not all farms fell neatly into one of two categories, of course, which Becker reminds us is usual and should be the basis of additional sociological insight.

5. Wendell Berry (1981, 132–133) makes the same point, though more elegantly and generalized to the whole of American agriculture.

6. Buttel, Larson, and Gillespie (1990) note the characterization of the problem of agriculture as the "diffusion and adoption of technological innovation." Their survey of the sociology of agriculture shows how research by sociologists largely followed the national political agenda of modernization and growth, at least until the 1970s.

7. I take the phrase from the dean of an influential agricultural school, delivered in a public lecture on the modernization of dairy farming.

8. The essays gathered by Comstock (1987) recognize a qualitative difference between family and factory farms. Lyson and Geisler (1992) describe environmental and social impacts of the shift from family to factory farms, a theme developed from Goldschmidt's study of the social impact of agribusiness in a California farm region during the 1940s. Vogeler (1981) carried this argument into the 1980s. Carol Bly's essay "Getting Tired" (1981, 8–13) puts the question into a comparison between the experience of bone-tiring agricultural work of the old system versus the droning alienation of unbroken hours at the controls of a huge combine.

Behind much of my thinking are the ideas of Piore and Sabel (1984) relating to the mixing of forms and scales of production.

Other sociologists who have asked similar questions include Cruise and Lyson (1991) and Davidson and Schwarzweller (1995) for the United States, Tovey (1982) for Ireland, and van der Ploeg (1985) for northern Italy.

9. Bill van Loo (1999, 9) describes the first commercial robotic milker installed on a farm in Ontario, Canada. The system uses computers, laser, and sensors: "When a cow steps into the milker which is located in a stall in the barn a board comes up and rests on her backside. Next the milker arm cleans the teats with rolls, these are then disinfected with sprayed on disinfectant. By now the computer has identified the cow and the laser finds the teats and the teat cups are attached. . . . As the cow is being milked a grain ration is being doled out. As milk flow ceases the milk cups are individually removed. Grain is stopped at the removal of the first teat cup. Meanwhile, cow ID, volume of milk per quarter, total milk volume, time of milking and conductivity of the milk (alerts for mastitis) have been measured and recorded. Any abnormalities or deviations from the norm are noted and flagged

on the computer printout." This extraordinary portrait either thrills an industrialist or evokes the sequences from Charlie Chaplin's *Modern Times*, when the automatic lunch ingester is tested on the workers.

10. My experiences with French and Swiss dairy farmers were brief but intense. In Switzerland I interviewed a farm family in the southern Alpine region who are neighbors of my colleague Ricabeth Steiger. Steiger subsequently researched agricultural statistics to determine that the family I studied was characteristic of the region, and of rural Switzerland generally. In the meantime Steiger did follow-up interviews with the farm family. In France my interviews were organized by Martine Duquesne, who located typical farms in the Normandy region. These interviews, which included a local agricultural agent and an artificial insemination specialist, as well as two farmers, were tape recorded, translated, and transcribed.

Finally, I note that not all farming in northern Europe follows this model. The industrial model of dairy farming is well advanced in Holland, and it was the Dutch after World War II who brought factory dairy farming to California and who presently dominate that industry.

11. Schwarzweller's edited collection, in press as this book goes to press, examines dairy farm restructuring worldwide and promises to be a long-needed comparative study.

12. Leopold 1949, 238–239.

13. Wojcik 1989, x.

References

Adams, Jane H. 1988. "The Decoupling of Farm and Household: Differential Consequences of Capitalist Development on Southern Illinois and Third World Family Farms." *Comparative Studies in Society and History* 30 (2): 453–482.

——. 1993. "Resistance to "Modernity": Southern Illinois Farm Women and the Cult of Domesticity." *American Ethnologist* 20 (1): 89–113.

——. 1994. *The Transformation of Rural Life: Southern Illinois, 1890–1990.* Chapel Hill: University of North Carolina Press.

Agee, James, and Walker Evans. 1939. *Let Us Now Praise Famous Men.* Cambridge, Mass.: Riverside Press.

Anderson, Nels. 1940. *Men on the Move.* Chicago: University of Chicago Press.

Bateson, Gregory, and Margaret Mead. 1942. *Balinese Character: A Photographic Analysis.* New York: New York Academy of Sciences.

Becker, Howard S. 1986. "Culture: A Sociological View." In *Doing Things Together,* 11–24. Evanston, Ill.: Northwestern University Press.

——. 1998. *Tricks of the Trade.* Chicago: University of Chicago Press.

Berry, Wendell. 1981. *The Gift of Good Land.* San Francisco: North Point Press.

Blumer, Herbert. 1969. "The Methodological Position of Symbolic Interactionism." In *Symbolic Interactionism: Perspective and Meaning,* 1–61. New York: Prentice Hall.

Blau, Peter. 1964. *Exchange and Power in Social Life.* New York: John Wiley.

Bly, Carol. 1981. "Getting Tired." In *Letters from the Country,* 8–13. New York: Penguin.

Blythe, Ronald. 1969. *Akenfield: Portrait of an English Village.* New York: Random House.

Boffin, Tessa, and Jean Fraser, eds. 1991. *Stolen Glances: Lesbians Take Photographs.* London: Pandora.

Bolton, Richard, ed. 1989. *The Contest of Meaning: Critical Histories of Photography.* Cambridge, Mass.: MIT Press.

Bright, Deborah. 1989. "Of Mother Nature and Marboro Men: An Inquiry into the Cultural Meanings of Landscape Photography." In *The Contest of Meaning: Critical Histories of Photography,* edited by Richard Bolton. Cambridge, Mass.: MIT Press.

Brunner, Ernst. 1996. *Ernst Brunner: Photographien, 1937–1962.* Zurich: Offizin.

Buttel, Frederick H., Olaf F. Larson, and Gilbert W. Gillespie, Jr. 1990. *The Sociology of Agriculture.* New York: Greenwood Press.

Bush, Corlann Gee. 1987. "'He Isn't Half So Cranky as He Used to Be': Agricultural Mechanization, Comparable Worth, and the Changing Family Farm." In *"To Toil the Livelong Day": America's Women at Work, 1780–1980,* edited by Carol Groneman and Mary Beth Norton, 213–232. Ithaca, N.Y.: Cornell University Press.

Caldwell, Erskine, and Margaret Bourke-White. 1937. *You Have Seen Their Faces.* New York: Modern Age Books.

Canine, Craig. 1995. *Dream Reaper.* Chicago: University of Chicago Press.

Cohen, David Steven. 1992. *The Dutch-American Farm.* New York: New York University Press.

Cohen, Joyce Tenneson. 1978. *In/Sights: Self-Portraits by Women.* Boston: Godine.

Collier, John, Jr. 1967. *Visual Anthropology: Photography as a Research Method.* New York: Holt, Rinehart and Winston.

Collier, John, Jr., and Malcolm Collier. 1986. *Visual Anthropology: Photography as a Research Method.* Rev. ed. Albuquerque: University of New Mexico Press.

Comstock, Gary, ed. 1987. *Is There a Moral Obligation to Save the Family Farm?* Ames, Iowa: Iowa State University Press.

Cruise, James, and Thomas A. Lyson. 1991. "Beyond the Farmgate: Factors Related to Agricultural Performance in Two Dairy Communities." *Rural Sociology* 56 (1): 41–55.

Curtis, James. 1989. *Mind's Eye, Mind's Truth: FSA Photography Reconsidered.* Philadelphia: Temple University Press.

Dasgupta, Satish Chandra. 1945. *The Cow in India.* Vol. 1. Calcutta: Khadi Pratisthan.

Davidson, Andrew, and Harry K. Schwarzweller. 1995. "Marginality and Uneven Development: The Decline of Dairying in Michigan's North Country." *Sociologia Ruralis* 35 (1): 40–66.

Davis, John. 1992. *Exchange.* Minneapolis: University of Minnesota Press.

DeVeaux, Scott. 1997. *The Birth of Bebop: A Social and Musical History.* Berkeley: University of California Press.

Dubofsky, Melvyn. 1969. *We Shall Be All: A History of the IWW.* New York: Quadrangle.

Durkheim, Émile. 1933. *The Division of Labor in Society.* London: Macmillan.

Elbert, Sara. 1987. "Amber Waves of Gain: Women's Work in New York Farm Families." In *"To Toil the Livelong Day": America's Women at Work, 1780–1980,* edited by Carol Groneman and Mary Beth Norton, 250–268. Ithaca, N.Y.: Cornell University Press.

Fink, Deborah. 1986. *Open Country, Iowa: Rural Women, Tradition and Change.* Albany, N.Y.: SUNY Press.

——. 1992. *Agrarian Women: Wives and Mothers in Rural Nebraska, 1880–1940.* Chapel Hill:University of North Carolina Press.

Firth, Raymond. 1967. *Themes in Economic Anthropology.* London: Tavistock Publications.

Fisher, Emily. 1985. *George Did It: A History of the Fisher Family.* Unpublished manuscript.

Fleischhauer, Carl, and Beverly W. Brannan, eds. 1980. *Documenting America: 1935–1943.* Berkeley: University of California Press.

Galusha, Diane. 1994. *Through a Woman's Eye: Pioneering Photographers in Rural Upstate.* Hensonville, N.Y.: Black Dome Press.

Ganzel, Bill. 1984. *Dust Bowl Descent.* Lincoln: University of Nebraska Press.

Gerth, Hans, and C. Wright Mills. 1946. *From Max Weber: Essays in Sociology.* New York: Oxford University Press.

Glaser, B., and A. Strauss. 1967. *The Discovery of Grounded Theory.* Chicago: Aldine.

Gold, Steven. 1991. "Ethnic Boundaries and Ethnic Entrepreneurship: A Photo-Elicitation Study." *Visual Sociology* 6 (2): 9–23.

Goldschmidt, Walter. 1978. *As You Sow: Three Studies in the Social Consequences of Agribusiness.* Montclair, N.J.: Allanheld, Osmun.

Gouldner, Alvin. 1960. "The Norm of Reciprocity." *American Sociological Review* 25 (2): 161–187.
Groneman, Carol, and Mary Beth Norton, eds. 1987. *"To Toil the Livelong Day": America's Women at Work, 1780–1980.* Ithaca, N.Y.: Cornell University Press.
Guillet, David. 1980. "Reciprocal Labor and Peripheral Capitalism in the Central Andes." *Ethnology* 19, no. 2 (April): 151–167.
Gunasinghe, Newton. 1976. "Social Change and the Disintegration of a Traditional System of Exchange Labour in Kandyan, Sri Lanka." *Sociological Bulletin* 25 (2): 168–184.
Harper, Douglas. 1987. *Working Knowledge: Skill and Community in a Small Shop.* Chicago: University of Chicago Press.
——. 1991. "On 'Methodological Monism' in Rural Sociology." *Rural Sociology* 56 (1): 70–88.
Homans, George C. 1958. "Social Behavior as Exchange." *American Journal of Sociology* 63:597–606.
Hostetler, John. 1980. *Amish Society.* Baltimore: Johns Hopkins University Press.
Huff, Charles. 1996. *New York State Dairy Statistics.* Albany: State of New York, Department of Agriculture and Markets, Division of Dairy Industry Services and Producer Security.
Ingold, Tim. 1983. "Gathering the Herds: Work and Co-operation in a Northern Finnish Community." *Ethos* 48 (3–4): 133–159.
Jellison, Katherine. 1993. *Entitled to Power: Farm Women and Technology, 1913–1963.* Chapel Hill: University of North Carolina Press.
Keller, Urich. 1985. *Highway as Habitat: A Roy Stryker Documentation, 1943–1955.* Seattle: University of Washington Press.
Klinkenborg, Verlyn. 1986. *Making Hay.* New York: Nick Lyons Books.
Lange, Dorothea, and Paul S. Taylor. 1939. *An American Exodus: A Record of Human Erosion.* New York: Modern Library.
Laxness, Halldor. [1935] 1997. *Independent People.* New York: Vintage.
Lee, Dorothy. N.d. *A History of Pillar Point.* Unpublished manuscript.
Lemann, Nicholas. 1983. *Out of the Forties.* Austin: Texas Monthly Press.
Leopold, Aldo. 1949. *A Sand County Almanac.* New York: Oxford University Press.
Liebow, Elliot. 1993. *Tell Them Who I Am.* New York: Free Press.
Lyson, T. A., and C. Geisler. 1992. "Toward a Second Agricultural Divide: The Restructuring of American Agriculture." *Sociologia Ruralis* 32 (2/3): 248–263.
Malinowski, Bronislaw. 1922. *Argonauts of the Western Pacific.* London: Routledge and Kegan Paul.
Mauss, Marcel. 1967. *The Gift.* New York: Norton.
McMurry, Sally. 1995. *Transforming Rural Life: Dairying Families and Agricultural Change, 1820–1885.* Baltimore: Johns Hopkins University Press.
Meyer, Katherine, and Linda Lobao. 1997. "Farm Couples and Crisis Politics: The Importance of Household, Spouse, and Gender in Responding to Economic Decline." *Journal of Marriage and the Family* 59 (February): 204–218.
Morrissey, James. N.d. *Reflections of the Tractor Salesman: Converting Horse Farmers into Tractor Operators.* Plymouth, Ind.: James Michael Morrissey.
Neumaier, D., ed. 1995. *Reframings: New Feminist Photographies.* Philadelphia: Temple University Press.
New York Agricultural Statistics. 1996–1997. Albany: New York Agricultural Statistics Service.
Ohrn, Karin Becker. 1980. *Dorothea Lange and the Documentary Tradition.* Baton Rouge: Louisiana State University Press.
Okrent, Daniel. 1989. *The Way we Were: New England Then, New England Now.* New York: Grove Weidenfield.

O'Neil, Brian Juan. 1982. "Cooperative Work in a Northern Portuguese Village." *Analise Social* 18, 1 (70): 7–34.

Osterud, Nancy Grey. 1991. *Bonds of Community: The Lives of Farm Women in Nineteenth-Century New York.* Ithaca, N.Y.: Cornell University Press.

Parker, Doris. 1985. "Echos from the Valley." *St. Lawrence County Historical Quarterly,* April, 13.

Perelman, Michael. 1976. "Efficiency in Agriculture: The Economics of Energy." In *Radical Agriculture,* edited by Richard Merrill. New York: Harper and Row.

Plattner, Steven W. 1983. *Roy Stryker: U.S.A., 1943–1950: The Standard Oil (New Jersey) Photography Project.* Austin: University of Texas Press.

Piore, M., and C. F. Sabel. 1984. *The Second Industrial Divide.* New York: Basic Books.

Rendleman, Edith Bradley. 1996. *All Anybody Ever Wanted of Me Was to Work: The Memoirs of Edith Bradley Rendlemen.* Edited by Jane Adams. Carbondale: Southern Illinois University Press.

Rölvaag, O. E. 1927. *Giants in the Earth.* New York: Harper and Brothers.

Sachs, Carolyn E. 1983. *The Invisible Farmers: Women in Agricultural Production.* Totowa, N.J.: Rowman and Allanheld.

Sahlins, Marshall. 1972. *Stone Age Economics.* Chicago: Aldine

Salamon, Sonja. 1992. *Prairie Patrimony: Family, Farming, and Community in the Midwest.* Chapel Hill: University of North Carolina Press.

Schwarzweller, Harry K., and Andrew Davidson. 2000. *A Focus on Dairying: Research in Rural Sociology and Development.* Vol. 8. Oxford: Elsevier Science.

Sloan, Eric. 1965. *Diary of an Early American Boy.* New York: Ballantine Books.

Solomon-Godeau, Abigail. 1991. *Photography at the Dock: Essays on Photographic History, Institutions and Practices.* Minneapolis: University of Minnesota Press.

Spencer, Dunstan S. C. 1979. "Labor Market Organization, Wage Rates and Employment in Rural Areas of Sierra Leone." *Labour and Society* 4, no. 3 (July):, 293–308.

Stadtfeld, Curtis K. 1972. *From the Land and Back.* New York: Charles Scribner's Sons.

Sterchi, Beat. 1988. *Cow.* New York: Pantheon.

Sterngold, James. 1999. "Dairy Farmers in California Find Gold in Urban Sprawl." *New York Times,* 22 October.

St. Lawrence County Agriculture. 1981. Canton, N.Y.: St. Lawrence Cooperative Extension Association.

Suchar, Charles S. 1988. "Photographing the Changing Material Culture of a Gentrified Community." *Visual Sociology Review* 3 (2): 17–22.

Tovey, H. 1982. "Milking the Farmer? Modernization and Marginalisation in Irish Dairy Farming." In *Power, Conflict and Inequality,* edited by Mary Kelly, Liam O'Dowd, and James Wickham, 68–89. Dublin: Turoe Press.

Trefil, James. 1997. "Nitrogen." *Smithsonian,* October, 70–78.

Tucker, Anne, ed. 1975. *The Woman's Eye.* New York: Knopf.

Ulrich, Laurel Thatcher. 1987. "Housewife and Gadder: Themes of Self-Sufficiency and Community in Eighteenth-Century New England." In *"To Toil the Livelong Day": America's Women at Work, 1780–1980,* edited by Carol Groneman and Mary Beth Norton, 21–34. Ithaca, N.Y.: Cornell University Press.

United States Census of Agriculture. 1925, 1945, 1987, 1997. Washington: United States Government Printing Office.

van der Does, Sonja Edelaar, Imke Gooskens, Margreet Liefting, and Marije van Mierlo. 1992. "Reading Images: A Study of a Dutch Neighborhood." *Visual Sociology* 7 (1): 4–68.

van der Ploeg, Jan Douwe. 1985. "Patterns of Farming Logic, Structuration of Labour and Impact of Externalization: Changing Dairy Farming in Northern Italy." *Sociologia Ruralis* 25 (1): 5–25.

van Loo, Bill. 1999. "Robotic Milkers? First Commercial Unit in Ontario." *Northern Adirondack Agricultural News*, September, 9.
Vogeler, Ingolf. 1981. *The Myth of the Family Farm: Agribusiness Dominance of U.S. Agriculture.* Boulder, Colo.: Westview Press.
Weitzman, David. 1991. *Thrashin' Time: Harvest Days in the Dakotas.* Boston: David R. Godine.
Wojcik, Jan. 1989. *Issues of Agriculture.* Purdue, Ind.: Purdue University Press.

Index